THE RATIONAL UNIFIED PROCESS

Philippe Kruchten

ADDISON-WESLEY

An imprint of Addison Wesley Longman, Inc.

Reading, Massachusetts • Harlow, England • Menlo Park, California
Berkeley, California • Don Mills, Ontario • Sydney
Bonn • Amsterdam • Tokyo • Mexico City

The publisher offers discounts on this book when ordered in quantity for special sales. For more information, please contact:

AWL Direct Sales
Addison Wesley Longman, Inc.
One Jacob Way
Reading, Massachusetts 01867

Library of Congress Cataloging-in-Publication Data
Kruchten, Philippe.
The Rational Unified Process : an introduction / Philippe Kruchten.
p. cm. — (Addison-Wesley Object Technology Series)
Includes bibliographical references and index.
ISBN 0-201-60459-0 (alk. paper)
1. Computer software—Development. 2. Software engineering.
I. Title. II. Series.
QA76.76.D47K78 1998
005.1--dc21 98-44897
CIP

Executive Editor: J. Carter Shanklin
Project Editor: Krysia Bebick
Production Coordinator: Marilyn Rash
Copyeditor: Betsy Hardinger
Composition: Bookwrights, Rockland, ME

ISBN 0-201-60459-0

Visit AWL on the Web: www.awl.com/cseng/
Text printed on recycled and acid-free paper.

1 2 3 4 5 6 7 8 9 10–CRS–02 01 00 99 98

First printing, November 1998

Contents

To
Sylvie, Alice, Zoé,
and Nicolas

Preface

THE RATIONAL UNIFIED PROCESS is a software engineering process developed and marketed by Rational Software. It provides a disciplined approach to assigning and managing tasks and responsibilities within a development organization. The goal of this process is to produce, within a predictable schedule and budget, high-quality software that meets the needs of its end users.

The Rational Unified Process captures many of the best practices in modern software development and presents them in a tailorable form that is suitable for a wide range of projects and organizations. The Rational Unified Process delivers these best practices to the project team online in a detailed, practical form.

This book provides an introduction to the concepts, structure, contents, and motivation of the Rational Unified Process.

GOALS OF THIS BOOK

In this book, you will

- Learn what the Rational Unified Process is and what it is not
- Master the vocabulary of the Rational Unified Process and understand its structure
- Develop an appreciation for the best practices that we have synthesized in this process
- Understand how the Rational Unified Process can give you the guidance you need for your specific responsibility in a project

This book is an integral part of the Rational Unified Process, but it is not the complete Rational Unified Process. Rather, it is only a small subset. In the full Rational Unified Process you will find the detailed guidance needed to carry out your work. See Chapter 2, Figure 2-2 for a complete map of all the manuals that constitute the Rational Unified Process. The full Rational Unified Process product—the online *knowledge base*—can be obtained from Rational Software.

This introductory book makes numerous references to the Unified Modeling Language (UML), but it is not an introduction to the UML. That is the focus of two other books: *The Unified Modeling Language User Guide* and *The Unified Modeling Language Reference Manual*.

This introductory book speaks about modeling and object-oriented techniques, but it is not a design method and it does not teach you how to model. Detailed steps and guidance on the various techniques that are embedded in the Rational Unified Process can be found only in the process product.

Several chapters of this book discuss project management issues. They describe aspects of planning an iterative development, managing risks, and so on. This book, however, is by no means a complete manual on project management and software economics. For more information, we refer you to the book *Software Project Management: A Unified Framework*.

The Rational Unified Process is a specific and detailed instance of a more generic process described in the textbook *The Unified Software Development Process*.

WHO SHOULD READ THIS BOOK?

The Rational Unified Process: An Introduction is written for a wide range of people involved in software development: project managers, developers, quality engineers, process engineers, method specialists, system engineers, and analysts.

This book is especially relevant to members of an organization that has adopted the Rational Unified Process or is about to adopt it. It is likely that an organization will tailor the Rational Unified Process to suit its needs, but the core process described in this book should remain the common denominator across all instances of the Rational Unified Process.

This book will be a useful companion to students taking one of the many professional education courses delivered by Rational Software and its partners from industry and academia. It provides a general context for the specific topics covered by the course.

This book assumes that you have a basic understanding of software development. It does not require any specific knowledge of a programming language, of an object-oriented method, or of the Unified Modeling Language.

HOW TO USE THIS BOOK

Software professionals who are working in an organization that has adopted the Rational Unified Process, in whole or part, should read the book linearly. The chapters have been organized in a natural progression.

Project managers can limit their reading to Chapters 1, 2, 4, and 7, which provide an introduction to the implications of an iterative, risk-driven development process.

Process engineers or methodologists may have to tailor and install the Rational Unified Process in their organizations. They should carefully study Chapters 3 and 17, which describe the process structure and the overall approach to implementing the Rational Unified Process.

ORGANIZATION AND SPECIAL FEATURES

The book has two parts.

Part I describes the process, its context, its history, its structure, and its software development life cycle. It describes some of the key features that differentiate the Rational Unified Process from other software development processes.

- Chapter 1: Software Development Best Practices
- Chapter 2: The Rational Unified Process
- Chapter 3: Static Structure: Process Representation

- Chapter 4: Dynamic Structure: Iterative Development
- Chapter 5: An Architecture-Centric process
- Chapter 6: A Use-Case-Driven Process

Part II gives an overview of the various process components, or workflows. There is one chapter for each core process workflow.

- Chapter 7: The Project Management Workflow
- Chapter 8: The Business Modeling Workflow
- Chapter 9: The Requirements Workflow
- Chapter 10: The Analysis and Design Workflow
- Chapter 11: The Implementation Workflow
- Chapter 12: The Test Workflow
- Chapter 13: The Configuration and Change Management Workflow
- Chapter 14: The Deployment Workflow
- Chapter 15: The Environment Workflow
- Chapter 16: Iteration Workflows
- Chapter 17: Implementing the Rational Unified Process

Most workflow chapters are organized into six sections:

- Purpose of the workflow
- Definitions and key concepts
- Workers and artifacts
- A typical workflow: an overview of the activities
- Tool support
- Summary

Two appendixes summarize all the workers (the "roles" of the process) and artifacts (the "work products" of the process) that are introduced in Chapters 7 through 15. A Glossary of common terms is provided, as is a short annotated Bibliography.

ACKNOWLEDGMENTS

The Rational Unified Process reflects the wisdom of a great many software professionals from Rational Software and elsewhere. The history of the process can be found in Chapter 2. But pulling out a book, even as small and modest as this one, required the dedicated effort of a slate of people, whom I would like to recognize here.

The members of the Rational Process Development Group assembled the Rational Unified Process 5.0 and contributed to this introduction. You will see some of their names associated with specific chapters.

- Maria Ericsson developed the business modeling and requirements management aspect and was a keeper of the process architecture.
- Stefan Bylund contributed to the analysis and design chapter and integrated the user-interface design aspects.
- Kurt Bittner contributed to the analysis and design chapter, contributed to project management, and developed the data engineering aspects.
- Håkan Dyrhage contributed many ideas to the organization and structure of the process and to its implementation and configuration and also coordinated the development of the online version.
- Jas Madhur contributed the ideas on configuration management and change management.
- Bruce Katz contributed the testing aspects of the process.
- Margaret Chan was responsible for the product integration, and for the assembly of most of the artwork in this book.
- Debbie Gray is the devoted administrative assistant of a team spread across nine time zones.

We are very grateful to Grady Booch for writing Chapter 1.

Per Kroll is the marketing manager for the Rational Unified Process, and Paer Jansson is its product manager. Christina Gisselberg and Eric Turesson designed and developed the online version. Stefan Ahlqvist developed the ideas on user-interface design.

The Rational Unified Process and this book benefited from the reviews and ideas of Dave Bernstein, Grady Booch, Geoff Clemm, Catherine Connor, Mike Devlin, Christian Ehrenborg (Dr. Use-case), Sam Guckenheimer, Björn Gustafsson, Ivar Jacobson, Ron Krubek, Dean Leffingwell, Andrew Lyons, Bruce Malasky, Roger Oberg, Gary Pollice, Leslee Probasco, Terri Quatrani, Walker Royce, Jim Rumbaugh, John Smith, and Brian White.

Special thanks go to our British friends, who have always had some special interest in the Rational process: Ian Gavin, Ian Spence, and Mike Tudball.

The Frenglish and the Sweglish were ironed out by Joy Hemphill and Pamela Clarke.

And finally many thanks to my editor, J. Carter Shanklin, as well as Krysia Bebick, Marilyn Rash and her team, and all the staff at Addison Wesley Longman for getting this book out as quickly as they did.

FOR MORE INFORMATION

Information about the Rational Unified Process, such as a data sheet and a downloadable demo version, can be obtained from Rational Software via the Internet at www.rational.com/rup_info/.

If you are already using the Rational Unified Process, additional information is available from the Rational Unified Process Resource Center, which has extra goodies, updates, and links to partners. The hyperlink to the Resource Center is in the online version.

Academic institutions can contact Rational Software for information on a special program for including the Rational Unified Process in the curriculum.

Philippe Kruchten
Vancouver, B.C. Canada

Part I
The Process

Chapter 1

Software Development Best Practices

by Grady Booch

THIS CHAPTER SURVEYS best practices for software development and establishes a context for the Rational Unified Process.

THE VALUE OF SOFTWARE

Software is the fuel on which modern businesses are run, governments rule, and societies become better connected. Software has helped us create, access, and visualize information in previously inconceivable ways and forms. Globally, the breathtaking pace of progress in software has helped drive the growth of the world's economy. On a more human scale, software-intensive products have helped cure the sick and have given voice to the speechless, mobility to the impaired, and opportunity to the less able. From all these perspectives, software is an indispensable part of our modern world.[1]

The good news for software professionals is that worldwide economies are becoming increasingly dependent on software. The

1. Grady Booch, "Leaving Kansas," *IEEE Software* 15(1) Jan.–Feb. 1998, pp. 32–35.

kinds of software-intensive systems that technology makes possible and society demands are expanding in size, complexity, distribution, and importance.

The bad news is that the expansion of these systems in size, complexity, distribution, and importance pushes the limits of what we in the software industry know how to develop. Trying to advance legacy systems to more modern technology brings its own set of technical and organizational problems. Compounding the problem is that businesses continue to demand increased productivity and improved quality with faster development and deployment. Additionally, the supply of qualified development personnel is not keeping pace with the demand.

The net result is that building and maintaining software is hard and getting harder; building quality software in a repeatable and predictable fashion is harder still.

SYMPTOMS AND ROOT CAUSES OF SOFTWARE DEVELOPMENT PROBLEMS

Different software development projects fail in different ways—and, unfortunately, too many of them fail—but it is possible to identify a number of common symptoms that characterize these kinds of projects:[2,3]

- Inaccurate understanding of end-user needs
- Inability to deal with changing requirements
- Modules that don't fit together
- Software that's hard to maintain or extend
- Late discovery of serious project flaws
- Poor software quality
- Unacceptable software performance

2. Caper Jones, *Patterns of Software Systems Failure and Success*. London: International Thompson Computer Press, 1996.

3. Edward Yourdon, *Death March: Managing "Mission Impossible" Projects*. Upper Saddle River, NJ: Prentice Hall, 1997.

- Team members in each other's way, making it impossible to reconstruct who changed what, when, where, and why
- An untrustworthy build-and-release process

Unfortunately, treating these symptoms does not treat the disease. For example, the late discovery of serious project flaws is only a symptom of larger problems, namely, subjective project status assessment and undetected inconsistencies in the project's requirements, designs, and implementations.

Although different projects fail in different ways, it appears that most of them fail because of a combination of the following root causes:

- Ad hoc requirements management
- Ambiguous and imprecise communication
- Brittle architectures
- Overwhelming complexity
- Undetected inconsistencies in requirements, designs, and implementations
- Insufficient testing
- Subjective project status assessment
- Failure to attack risk
- Uncontrolled change propagation
- Insufficient automation

SOFTWARE BEST PRACTICES

If you treat these root causes, not only will you eliminate the symptoms, but you'll also be in a much better position to develop and maintain quality software in a repeatable and predictable fashion.

That's what software best practices are all about: commercially proven approaches to software development that, when used in combination, strike at the root causes of software development problems.[4] They are called "best practices" not so much because

---

4. See the Software Program Manager's Network best practices work at http://www.spmn.com.

you can precisely quantify their value but rather because they are observed to be commonly used in industry by successful organizations. These best practices are as follows.

1. Develop software iteratively.
2. Manage requirements.
3. Use component-based architectures.
4. Visually model software.
5. Verify software quality.
6. Control changes to software.

DEVELOP SOFTWARE ITERATIVELY

Classic software development processes follow the waterfall life cycle, as illustrated in Figure 1-1. In this approach, development proceeds linearly from requirements analysis through design, code and unit testing, subsystem testing, and system testing.

The fundamental problem of this approach is that it pushes risk forward in time so that it's costly to undo mistakes from earlier phases. An initial design will likely be flawed with respect to its key requirements, and, furthermore, the late discovery of design de-

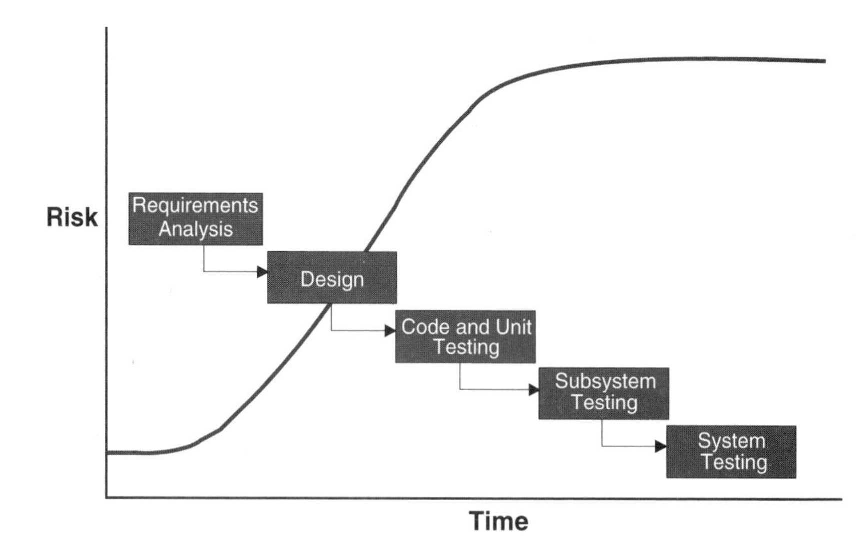

FIGURE 1-1 *The waterfall life cycle*

fects tends to result in costly overruns or project cancellation. As Tom Gilb aptly said, "If you do not actively attack the risks in your project, they will actively attack you."[5] The waterfall approach tends to mask the real risks to a project until it is too late to do anything meaningful about them.

An alternative to the waterfall approach is the iterative and incremental process, as shown in Figure 1-2.

In this approach, building on the work of Barry Boehm's spiral model,[6] the identification of risks to a project is forced early in the life cycle, when it's possible to attack and react to them in a timely and efficient manner.

This approach is one of continuous discovery, invention, and implementation, with each iteration forcing the development team

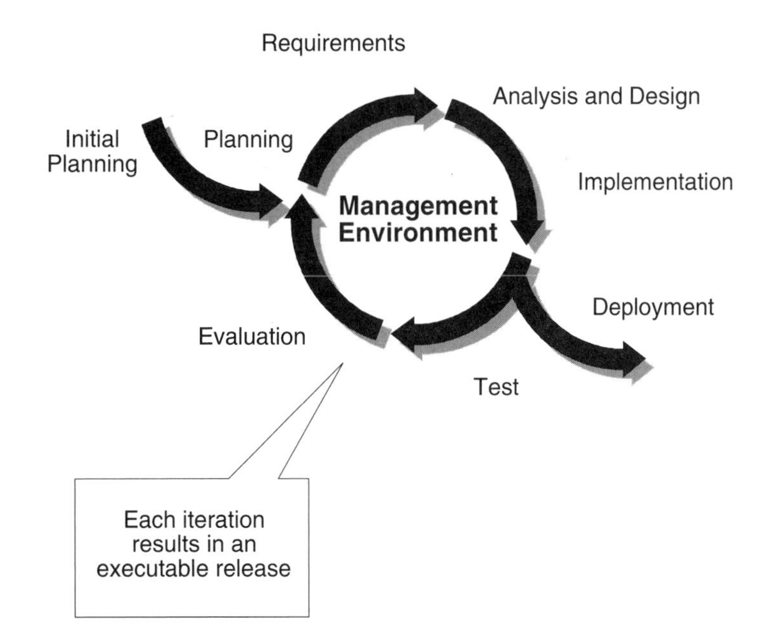

FIGURE 1-2 ***An iterative and incremental process***

5. Tom Gilb, *Principles of Software Engineering Management*. Harlow, UK: Addison-Wesley, 1988, p. 73.

6. Barry W. Boehm, "A Spiral Model of Software Development and Enhancement," *IEEE Computer*, May 1988, pp. 61–72.

to drive the project's artifacts to closure in a predictable and repeatable way.

Developing software iteratively offers a number of solutions to the root causes of software development problems.

1. Serious misunderstandings are made evident early in the life cycle, when it's possible to react to them.
2. This approach enables and encourages user feedback so as to elicit the system's real requirements.
3. The development team is forced to focus on those issues that are most critical to the project and are shielded from those issues that distract them from the project's real risks.
4. Continuous, iterative testing enables an objective assessment of the project's status.
5. Inconsistencies among requirements, designs, and implementations are detected early.
6. The workload of the team, especially the testing team, is spread out more evenly throughout the life cycle.
7. The team can leverage lessons learned and therefore can continuously improve the process.
8. Stakeholders in the project can be given concrete evidence of the project's status throughout the life cycle.

MANAGE REQUIREMENTS

The challenge of managing the requirements of a software-intensive system is that they are dynamic: you must expect them to change during the life of a software project. Furthermore, identifying a system's true requirements—those that weigh most heavily on the system's economic and technical goals—is a continuous process. Except for the most trivial system, it is impossible to completely and exhaustively state a system's requirements before the start of development. Indeed, the presence of a new or evolving system changes a user's understanding of the system's requirements.

A *requirement* is a condition or capability a system must meet. The active management of requirements encompasses three activities: eliciting, organizing, and documenting the system's required functionality and constraints; evaluating changes to these requirements and assessing their impact; and tracking and documenting trade-offs and decisions.

Managing the requirements of your project offers a number of solutions to the root causes of software development problems.

1. A disciplined approach is built into requirements management.
2. Communications are based on defined requirements.
3. Requirements can be prioritized, filtered, and traced.
4. An objective assessment of functionality and performance is possible.
5. Inconsistencies are more easily detected.
6. With suitable tool support, it is possible to provide a repository for a system's requirements, attributes, and traces, with automatic links to external documents.

USE COMPONENT-BASED ARCHITECTURES

Visualizing, specifying, constructing, and documenting a software-intensive system demand that the system be viewed from a number of different perspectives. Each of the different stakeholders—end users, analysts, developers, system integrators, testers, technical writers, and project managers—brings a different agenda to a project, and each of them looks at that system in a different way at different times over the project's life. A system's architecture is perhaps the most important deliverable that can be used to manage these different viewpoints and thereby controls the iterative and incremental development of a system throughout its life cycle.

A system's architecture encompasses the set of significant decisions about

- The organization of a software system
- The selection of the structural elements and their interfaces by which the system is composed
- Their behavior, as specified by the collaborations among those elements
- The composition of these structural and behavioral elements into progressively larger subsystems

- The architectural style that guides this organization: these elements and their interfaces, their collaborations, and their composition

Software architecture is concerned not only with structure and behavior but also with usage, functionality, performance, resilience, reuse, comprehensibility, economic and technology constraints and trade-offs, and aesthetic concerns.

Building resilient architectures is important because they enable economically significant degrees of reuse, offer a clear division of work among teams of developers, isolate hardware and software dependencies that may be subject to change, and improve maintainability.

Component-based development (CBD) is an important approach to software architecture because it enables the reuse or customization of existing components from thousands of commercially available sources. Microsoft's component object model (COM), the Object Management Group's (OMG) Common Object Request Broker Architecture (CORBA), and Sun's Enterprise JavaBeans (EJB) offer pervasive and widely supported platforms on which component-based architectures are made possible. As Figure 1-3 indicates, components make reuse possible on a larger scale, enabling systems to be composed from existing parts, off-the-shelf third-party parts, and a few new parts that address the specific domain and glue the other parts together.

Coupled with the practice of developing software iteratively, using component-based architectures involves the continuous evolution of a system's architecture. Each iteration produces an executable architecture that can be measured, tested, and evaluated against the system's requirements. This approach permits the team to continuously attack the most important risks to the project.

Using component-based architectures offers a number of solutions to the root causes of software development problems.

1. Components facilitate resilient architectures.
2. Modularity enables a clear separation of concerns among elements of a system that are subject to change.
3. Reuse is facilitated by leveraging standardized frameworks (such as COM+, CORBA, and EJB) and commercially available components.

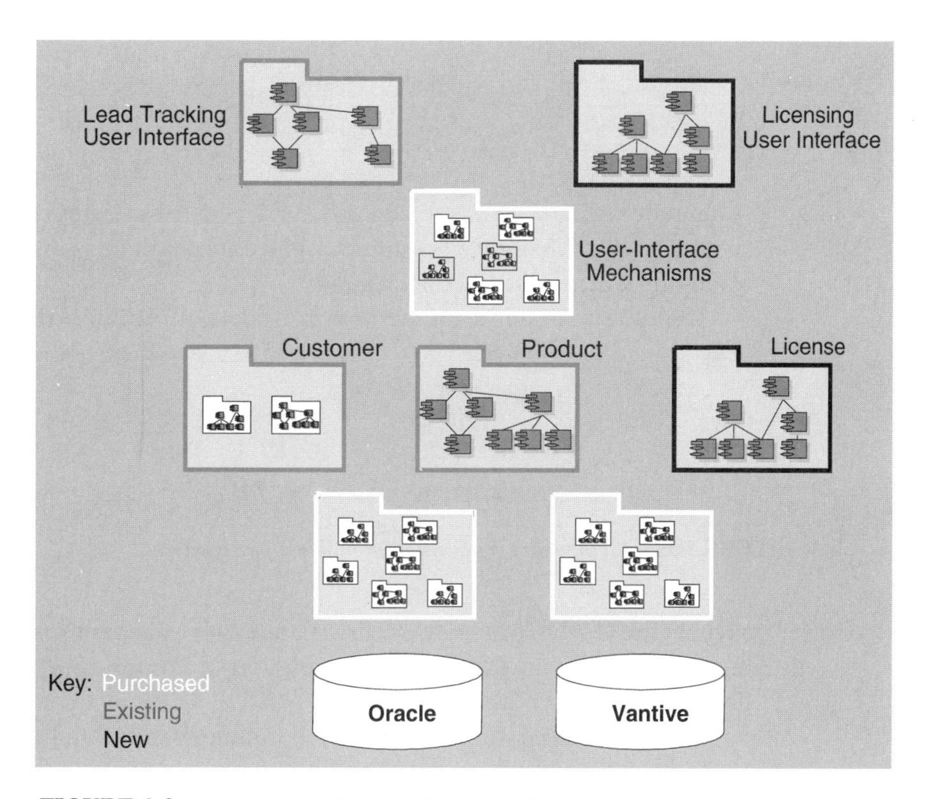

FIGURE 1-3 *A component-based software architecture*

4. Components provide a natural basis for configuration management.
5. Visual modeling tools provide automation for component-based development.

VISUALLY MODEL SOFTWARE

A model is a simplification of reality that completely describes a system from a particular perspective, as shown in Figure 1-4. We build models so that we can better understand the system we are modeling; we build models of complex systems because we cannot comprehend such systems in their entirety.

Modeling is important because it helps the development team visualize, specify, construct, and document the structure and behavior of a system's architecture. Using a standard modeling language

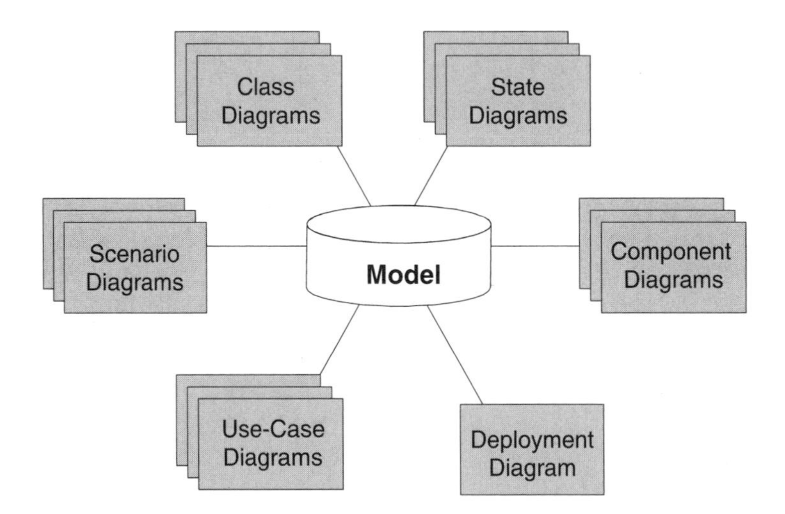

FIGURE 1-4 *Modeling a system from different perspectives*

such as the UML (Unified Modeling Language), different members of the development team can unambiguously communicate their decisions to one another.

Visual modeling tools facilitate the management of these models, letting you hide or expose details as necessary. Visual modeling also helps to maintain consistency between a system's artifacts: its requirements, designs, and implementations. In short, visual modeling helps improve a team's ability to manage software complexity.

When coupled with the practice of developing software iteratively, visual modeling helps you expose and assess architectural changes and communicate those changes to the entire development team. With the right kind of tools, you can then synchronize your models and source code during each iteration.

Modeling your software visually offers a number of solutions to the root causes of software development problems.

1. Use cases and scenarios unambiguously specify behavior.
2. Models unambiguously capture software design.
3. Nonmodular and inflexible architectures are exposed.
4. Detail can be hidden when necessary.
5. Unambiguous designs reveal their inconsistencies more readily.

6. Application quality starts with good design.
7. Visual modeling tools provide support for UML modeling.

VERIFY SOFTWARE QUALITY

As Figure 1-5 illustrates, software problems are 100 to 1,000 times more expensive to find and repair after deployment. For this reason, it's important to continuously assess the quality of a system with respect to its functionality, reliability, application performance, and system performance.

Verifying a system's functionality—the bulk of the testing activity—involves creating tests for each key scenario, each of which represents some aspect of the system's desired behavior. You can assess a system's functionality by asking which scenarios failed and where, as well as which scenarios and corresponding code have not yet been exercised. As you are developing your software iteratively, you test at every iteration, a process of continuous, quantitative assessment.

Verifying software quality offers a number of solutions (see next page) to the root causes of software development problems.

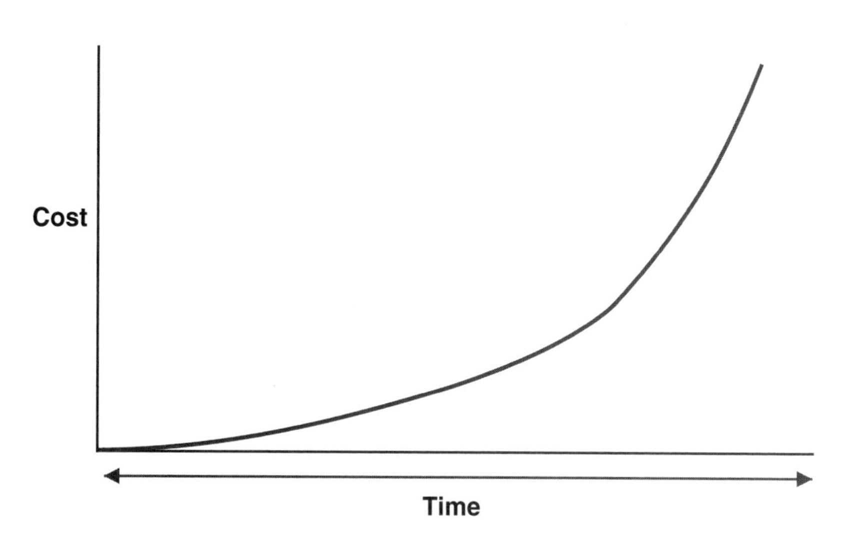

FIGURE 1-5 *The cost of fixing problems*

1. Project status assessment is made objective, and not subjective, because test results, and not paper documents, are evaluated.
2. This objective assessment exposes inconsistencies in requirements, designs, and implementations.
3. Testing and verification are focused on areas of highest risk, thereby increasing their quality and effectiveness.
4. Defects are identified earlier, radically reducing the cost of fixing them.
5. Automated testing tools provide testing for functionality, reliability, and performance.

CONTROL CHANGES TO SOFTWARE

A key challenge when you're developing software-intensive systems is that you must cope with multiple developers organized into different teams, possibly at different sites, working together on multiple iterations, releases, products, and platforms. In the absence of disciplined control, the development process rapidly degenerates into chaos.

Coordinating the activities and the artifacts of developers and teams involves establishing repeatable workflows for managing changes to software and other development artifacts. This coordination allows a better allocation of resources based on the project's priorities and risks, and it actively manages the work on those changes across iterations. Coupled with developing your software iteratively, this practice lets you continuously monitor changes so that you can actively discover and then react to problems.

Coordinating iterations and releases involves establishing and releasing a tested baseline at the completion of each iteration. Maintaining traceability among the elements of each release and among elements across multiple, parallel releases is essential for assessing and actively managing the impact of change.

Controlling changes to software offers a number of solutions to the root causes of software development problems.

1. The workflow of requirements change is defined and repeatable.

2. Change requests facilitate clear communications.
3. Isolated workspaces reduce interference among team members working in parallel.
4. Change rate statistics provide good metrics for objectively assessing project status.
5. Workspaces contain all artifacts, facilitating consistency.
6. Change propagation is assessable and controlled.
7. Changes can be maintained in a robust, customizable system.

THE RATIONAL UNIFIED PROCESS

A software development process has four roles.[7]

1. Provide guidance as to the order of a team's activities.
2. Specify which artifacts should be developed and when they should be developed.
3. Direct the tasks of individual developers and the team as a whole.
4. Offer criteria for monitoring and measuring the project's products and activities.

Without a well-defined process, your development team will develop in an ad hoc manner, with success relying on the heroic efforts of a few dedicated individual contributors. This is not a sustainable condition.

By contrast, mature organizations employing a well-defined process can develop complex systems in a repeatable and predictable way. Not only is that a sustainable business, but also it's one that can improve with each new project, thereby increasing the efficiency and productivity of the organization as a whole.

Such a well-defined process enables and encourages all of the best practices described earlier. When you codify these practices into a process, your development team can build on the collective experience of thousands of successful projects.

7. Grady Booch, *Object Solutions—Managing the Object Oriented Project*. Reading, MA: Addison-Wesley, 1995.

The Rational Unified Process, as described in the following chapters, builds on these six commercial best practices to deliver a well-defined process. This is the context for the Rational Unified Process, a software development process focused on ensuring the production of quality systems in a repeatable and predictable fashion.

SUMMARY

- Building quality software in a repeatable and predictable fashion is hard.
- There are a number of symptoms of common software development problems, and these symptoms are the observable results of deeper root causes.
- Six commercial best practices strike at the root causes of these software development problems.
  - Develop software iteratively.
  - Manage requirements.
  - Use component-based architectures.
  - Visually model software.
  - Verify software quality.
  - Control changes to software.
- The Rational Unified Process brings these best practices together in a form that is suitable for a wide range of projects and organizations.

Chapter 2
The Rational Unified Process

THIS CHAPTER GIVES AN OVERVIEW of the Rational Unified Process, introduces the process structure, describes the process product, and outlines its main features.

WHAT IS THE RATIONAL UNIFIED PROCESS?

The Rational Unified Process is a *software engineering process*. It provides a disciplined approach to assigning tasks and responsibilities within a development organization. Its goal is to ensure the production of high-quality software that meets the needs of its end users within a predictable schedule and budget.

The Rational Unified Process is a *process product*. It is developed and maintained by Rational Software and integrated with its suite of software development tools. It is available from Rational Software on CD-ROM or through the Internet. This book is an integral part of the Rational Unified Process but represents only a small fraction of the Rational Unified Process knowledge base. Later in this chapter we describe the physical structure of the process product.

The Rational Unified Process has a *process framework* that can be adapted and extended to suit the needs of an adopting organization. Later in this chapter we describe the logical structure of this process framework. Chapter 3 describes in more detail how the process framework is organized and introduces the *process model,* the elements that compose the process framework.

The Rational Unified Process captures many of the *best practices* in modern software development in a form that is suitable for a wide range of projects and organizations. In particular, it covers the six practices introduced in Chapter 1.

1. Develop software iteratively.
2. Manage requirements.
3. Use component-based architectures.
4. Visually model software.
5. Verify software quality.
6. Control changes to software.

THE RATIONAL UNIFIED PROCESS AS A PRODUCT

Many organizations have slowly become aware of the importance of a well-defined and well-documented software development process to the success of their software projects. Over the years, they have collected their knowledge and shared it with their developers. This collective know-how often grows out of methods, published textbooks, training programs, and small how-to notes amassed over several projects. Unfortunately, these practices often end up gathering dust in nice binders on a developer's shelf—rarely updated, rapidly becoming obsolete, and almost never followed.

In contrast, the Rational Unified Process has been developed as a product and is maintained as a product or a software tool is. The following are some of the key features of the Rational Unified Process.

- Regular upgrades are released by Rational Software.
- It is available online using Web technology, so it is literally at the fingertips of the developers.
- It can be tailored to suit an organization's needs.
- It is integrated with many of the software development tools in the Rational suite so that developers can access process guidance within the tool they are using.

This approach provides the following benefits:

- Distribution of the latest version to all project members on an intranet
- Direct availability of key information about the process through the use of index and search engine process guidance or policies, including the latest document templates
- Navigation from one part of the process to another, eventually branching out to a tool or to an external reference or guideline document such as the UML
- Easy inclusion of local, project-specific process improvements or special procedures
- Version control and management of variants of the process by each project or department

This online Rational Unified Process product gives you benefits that are difficult to achieve with a process that is available only in the form of a book or binder.

Structure of the Process Product

The Rational Unified Process product consists of the following.

1. An *online version* of the Rational Unified Process: a fully hyperlinked Web site description in HTML
   - Tool mentors providing additional process guidance when you're working with the Rational suite of software development tools, such as Rational Rose for visual modeling or ClearQuest for configuration management
   - Templates for all major process artifacts:
     - Rational SoDA templates, which help automate software documentation
     - RequisitePro templates, which help manage requirements
     - Microsoft Word templates for most of the documents
     - Microsoft Project templates for planning a process
     - Microsoft FrontPage templates for extending the online process itself
2. A *set of manuals* that describe the process

A printed set of manuals is useful for newcomers to the project, for people who want to concentrate on one aspect, or those who just want to read it on the bus. The introductory book you have in your hands is Volume 1 of this set.

The Rational Unified Process Online

The online process makes it possible to view the contents of the manuals with any one of the popular Web browsers and support frameworks, such as Netscape Navigator™ and Microsoft Internet Explorer™. The Rational Unified Process online has many hypertext links and interactive images that enable easy information navigation.

The Rational Unified Process online makes it easy to find the specific information you are looking for, thanks to

- Extensive hyperlinking
- Graphical navigation
- A hierarchical tree browser
- An exhaustive index
- A search engine

You can locate these facilities as shown in Figure 2-1, which is a snapshot of a process online page.

Rational Unified Process Manuals

Each of the nine manuals has its own purpose but is also closely related to each of the others. Cross-references are provided, making it easy to find the information necessary to finish your current tasks.

This book explains what you need to know before you start to work with the Rational Unified Process.

Rational Unified Process: Process Manual (see Figure 2-2) provides the detailed, step-by-step guidance that you need to develop successful software products. It explains what to do, how and when to do it, and who does it.

Rational Unified Process: Artifacts contains descriptions of the elements of information that are the input or output of the process. It explains how this information can be packaged in documents and reports.

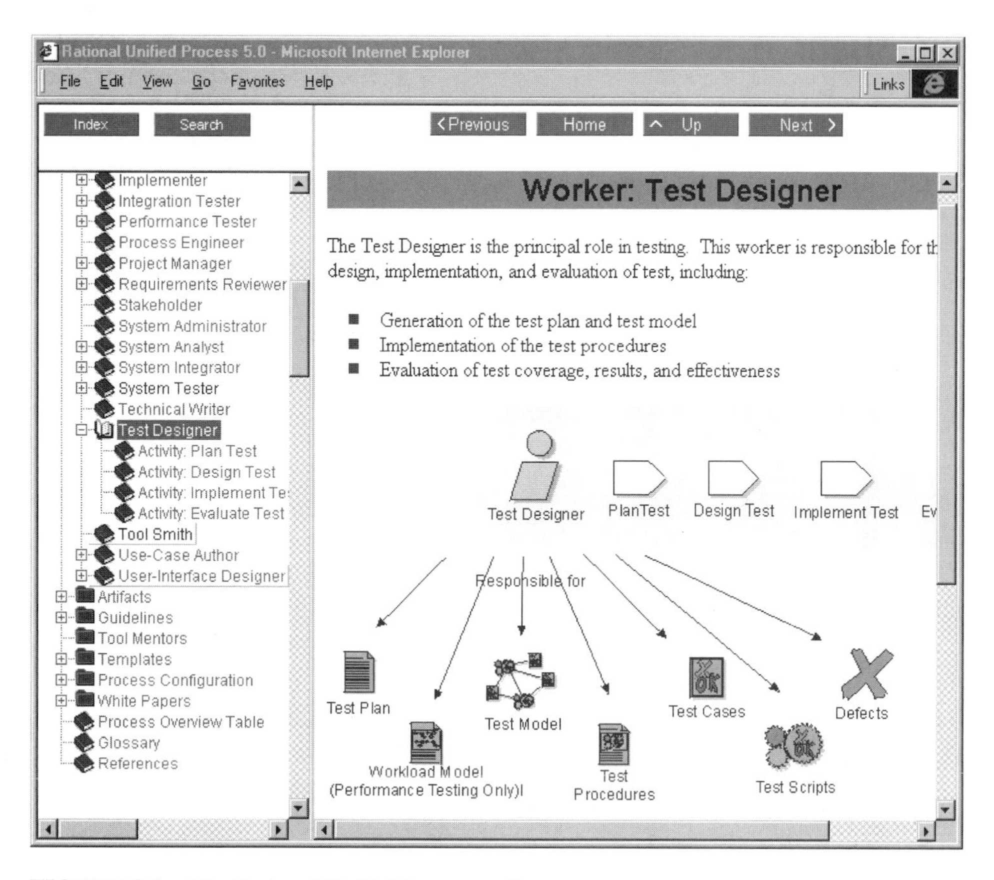

FIGURE 2-1 ***The Rational Unified Process online***

Rational Unified Process: Guidelines is a companion manual to the process manual and gives detailed advice on modeling elements: form, identification, quality, and other topics such as user-interface style.

Rational Unified Process: Process Configuration explains how to adapt the process to a specific project. To adapt the process, you decide which parts of the process (activities, modeling elements, or documents) to use and which individuals will do which activities.

Rational Unified Process: Project Management gives practical guidelines on planning, staffing, organizing, and controlling projects. It is separate from the core process manual because it addresses the concerns of different kinds of workers, such as managers and planners.

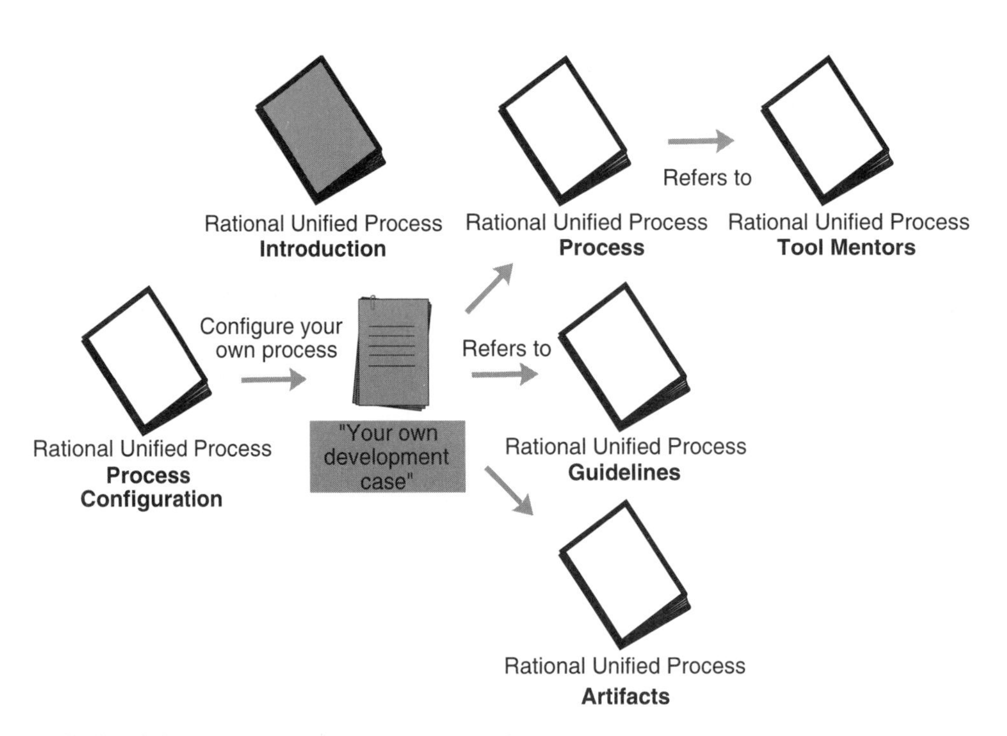

FIGURE 2-2 ***Rational Unified Process manuals***

Rational Unified Process: Business Modeling separately documents the business modeling workflow. It is apart from the core process manuals because business modeling is a separate process involving workers other than software developers.

The final two manuals—*Rational Unified Process: Ada Programming Guidelines* and *Rational Unified Process: C++ Programming Guidelines*—are booklets containing programming guidelines for Ada and for C++, respectively.

PROCESS STRUCTURE: TWO DIMENSIONS

Figure 2-3 shows the overall *architecture* of the Rational Unified Process. The process has two structures or, if you prefer, two dimensions.

- The horizontal axis represents time and shows the life-cycle aspects of the process as it unfolds.
- The vertical axis represents core process workflows, which group activities logically by nature.

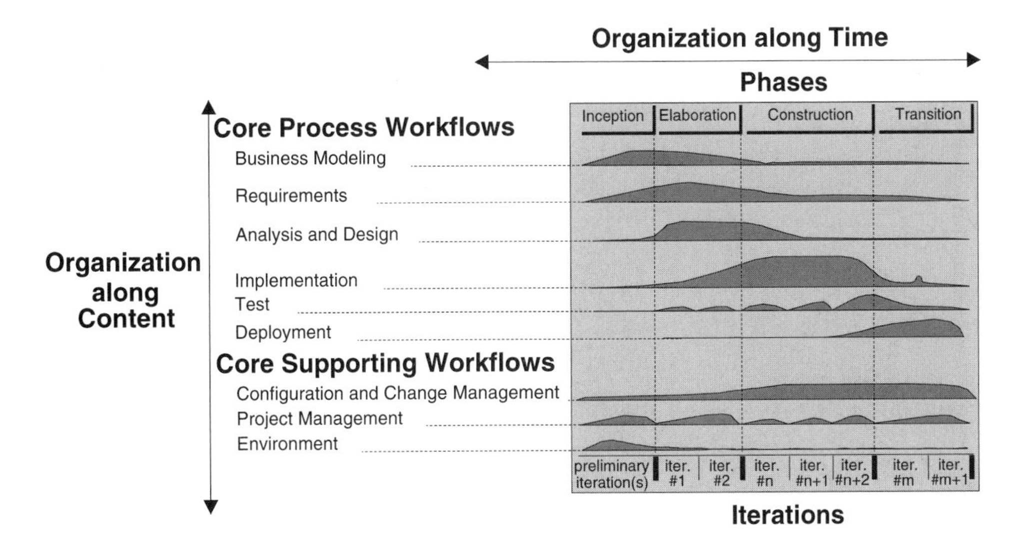

FIGURE 2-2 *Rational Unified Process manuals*

The first dimension represents the dynamic aspect of the process as it is enacted, and it is expressed in terms of cycles, phases, iterations, and milestones. This is further described in Chapter 4, Dynamic Structure: Iterative Development.

The second dimension represents the static aspect of the process: how it is described in terms of process components, activities, workflows, artifacts, and workers. This is further described in Chapter 3, Static Structure: Process Representation.

SOFTWARE BEST PRACTICES

This section revisits the six core best practices introduced in Chapter 1 and maps them to components of the Rational Unified Process.

Iterative Development

The iterative approach recommended by the Rational Unified Process is generally superior to a linear, or waterfall, approach for a number of reasons.

- It lets you take into account changing requirements. The truth is that requirements usually change. Requirements change and requirements "creep" have always been primary sources of project trouble, leading to late delivery, missed schedules, unsatisfied customers, and frustrated developers.

- In the Rational Unified Process, integration is not one "big bang" at the end; instead, elements are integrated progressively. This iterative approach is almost a process of continuous integration. What used to be a lengthy time of uncertainty and pain—taking up to 40% of the total effort at the end of a project—is now broken down into six to nine smaller integrations that begin with far fewer elements to integrate.

- The iterative approach lets you mitigate risks earlier because integration is generally the only time that risks are discovered or addressed. As you unroll the early iterations you go through all process components, exercising many aspects of the project, such as tools, off-the-shelf software, people skills, and so on. Perceived risks will prove not to be risks, and new, unsuspected risks will be revealed.

- It provides management with a means of making tactical changes to the product—for example, to compete with existing products. You can decide to release a product early with reduced functionality to counter a move by a competitor, or you can adopt another vendor for a given technology.

- It facilitates reuse because it is easier to identify common parts as they are partially designed or implemented instead of identifying all commonality in the beginning. Identifying and developing reusable parts is difficult. Design reviews in early iterations allow architects to identify unsuspected potential reuse and then develop and mature common code for it in subsequent iterations.

- It results in a more robust architecture because you correct errors over several iterations. Flaws are detected even in the early iterations as the product moves beyond inception into elaboration, and not in a massive testing phase at the end. Performance bottlenecks are discovered at a time when they can still be addressed instead of creating panic on the eve of delivery.

- Developers can learn along the way, and their various abilities and specialties are more fully employed during the entire life cycle. Testers start testing early, technical writers write early, and so on. In a non-iterative development, the same people would be waiting around to begin their work, making plan after plan but not making any concrete progress. What can a tester test when the product consists only of three feet of design documentation on a shelf? Training needs or the need for additional people is spotted early, during assessment reviews.

- The development process itself can be improved and refined along the way. The assessment at the end of an iteration not only looks at the status of the project from a product/schedule perspective but also analyzes what should be changed in the organization and in the process to make it perform better in the next iteration.

Project managers often resist the iterative approach, seeing it as a kind of endless and uncontrolled hacking. In the Rational Unified Process, the iterative approach is very controlled; iterations are planned in number, duration, and objectives. The tasks and responsibilities of the participants are defined. Objective measures of progress are captured. Some reworking takes place from one iteration to the next, but this, too, is carefully controlled.

Chapter 4 describes this iterative approach in more detail, and Chapter 7 describes how to manage an iterative process and, in particular, how to plan it.

Requirements Management

Requirements management is a systematic approach to eliciting, organizing, communicating, and managing the changing requirements of a software-intensive system or application.

The benefits of effective requirements management include the following.

- *Better control of complex projects*
  Lack of understanding of the intended system behavior as well as requirements creep are common factors in out-of-control projects.
- *Improved software quality and customer satisfaction*
  The fundamental measure of quality is whether a system does what it is supposed to do. This can be assessed only

when all stakeholders[1] have a common understanding of what must be built and tested.

- *Reduced project costs and delays*
  Fixing errors in requirements is very expensive; therefore, decreasing these errors early in the development cycle cuts project costs and prevents delays.
- *Improved team communication*
  Requirements management facilitates the involvement of users early in the process, helping to ensure that the application meets their needs. Well-managed requirements build a common understanding of the project needs and commitments among the stakeholders: users, customers, management, designers, and testers.

In Chapter 9, The Requirements Workflow, we revisit and expand on this important feature of the Rational Unified Process. Chapter 13 discusses the aspects related to tracking changes.

Architecture and Use of Components

Use cases drive the Rational Unified Process throughout the entire life cycle, but the design activities are centered on the notion of *architecture*—either system architecture or, for software-intensive systems, software architecture. The main focus of the early iterations of the process is to produce and validate a software architecture that, in the initial development cycle, takes the form of an executable architectural prototype that gradually evolves to become the final system in later iterations.

The Rational Unified Process provides a methodical, systematic way to design, develop, and validate an architecture. It offers templates for describing an architecture based on the concept of multiple architectural views. It provides for the capture of architectural style, design rules, and constraints. The design process component contains specific activities aimed at identifying archi-

---

1. A stakeholder is any person or representative of an organization who has a stake—a vested interest—in the outcome of a project. A stakeholder can be an end user, a purchaser, a contractor, a developer, a project manager, or anyone else who cares enough or whose needs must be met by the project.

tectural constraints and architecturally significant elements as well as guidelines on how to make architectural choices. The management process shows how planning the early iterations takes into account the design of an architecture and the resolution of the major technical risks.

A software *component* can be defined as a nontrivial piece of software, a module, a package, or a subsystem that fulfills a clear function, has a clear boundary, and can be integrated into a well-defined architecture. It is the physical realization of an abstraction in your design. Component-based development can take various flavors.

- In defining a modular architecture, you identify, isolate, design, develop, and test well-formed components. These components can be individually tested and gradually integrated to form the whole system.
- Furthermore, some of these components can be developed to be reusable, especially the components that provide common solutions to a wide range of common problems. These reusable components are typically larger than mere collections of utilities or class libraries. They form the basis of reuse within an organization, increasing overall software productivity and quality.
- More recently, the advent of commercially successful infrastructures supporting the concept of software component—such as Common Object Request Broker Architecture (CORBA), the Internet, ActiveX, and JavaBeans—has launched a whole industry of off-the-shelf components for various domains, allowing developers to buy and integrate components rather than develop them in-house.

The first point exploits the old concepts of modularity and encapsulation, bringing the concepts underlying object-oriented technology a step further. The final two points shift software development from programming software (one line at a time) to composing software (by assembling components).

The Rational Unified Process supports component-based development in several ways.

- The iterative approach allows developers to progressively identify components and decide which ones to develop, which ones to reuse, and which ones to buy.
- The focus on software architecture allows you to articulate the structure. The architecture enumerates the components and the ways they integrate as well as the fundamental mechanisms and patterns by which they interact.
- Concepts such as packages, subsystems, and layers are used during analysis and design to organize components and specify interfaces.
- Testing is organized around single components first and then is gradually expanded to include larger sets of integrated components.

Chapter 5 defines and expands on the concept of architecture and its central role in the Rational Unified Process.

Modeling and the UML

A large part of the Rational Unified Process is about developing and maintaining *models* of the system under development. Models help us to understand and shape both the problem and its solution. A model is a simplification of the reality that helps us master a large, complex system that cannot be comprehended in its entirety. We introduce several models in this book: use-case model (Chapter 6), business models (Chapter 8), design model and analysis model (Chapter 10), and test model (Chapter 12).

The Unified Modeling Language (UML) is a graphical language for visualizing, specifying, constructing, and documenting the artifacts of a software-intensive system. The UML gives you a standard means of writing the system's blueprints, covering conceptual items such as business processes and system functions as well as concrete items, such as classes written in a specific programming language, database schemas, and reusable software components.[2]

The UML is a common language to express the various models, but it does not tell you how to develop software. It provides the vo-

2. Grady Booch et al., *UML Users Guide*. Reading, MA: Addison Wesley Longman, 1998.

cabulary, but it doesn't tell you how to write the book. That is why Rational has developed the Rational Unified Process hand-in-hand with the UML to complement our work with the UML. The Rational Unified Process is a guide to the effective use of the UML for modeling. It describes which model you need, why you need it, and how to construct it.

Quality of Process and Product

Often people ask why there is no worker in charge of quality in the Rational Unified Process. The answer is that quality is not added to a product by a few persons. Instead, quality is the responsibility of every member of the development organization. In software development, our concern about quality is focused on two areas: product quality and process quality.

- *Product quality*
  The quality of the principal product being produced (the software or system) and all the elements it comprises (for example, components, subsystems, architecture, and so on).
- *Process quality*
  The degree to which an acceptable process (including measurements and criteria for quality) was implemented and adhered to during the manufacturing of the product. Additionally, process quality is concerned with the quality of the artifacts (such as iteration plans, test plans, use-case realizations, design model, and so on) produced in support of the principal product.

The Rational Unified Process, however, focuses on verifying and objectively assessing whether or not the product meets the expected level of quality. This is the primary purpose of the test workflow presented in Chapter 12 as well as other process and products metrics.

Configuration and Change Management

Particularly in an iterative development, many work products are often modified. By allowing flexibility in the planning and execution of the development and by allowing the requirements to evolve, iterative development emphasizes the vital issues of keeping track of changes and ensuring that everything and everyone is in sync. Focused closely on the needs of the development organization, change management is a systematic approach to managing changes in requirements, design, and implementation. It also covers the important activities of keeping track of defects, misunderstandings, and project commitments as well as associating these activities with specific artifacts and releases. Change management is tied to configuration management and to measurements.

Chapter 13, The Configuration and Change Management Workflow, expands on these important aspects of software management and their interrelationships.

OTHER KEY FEATURES OF THE RATIONAL UNIFIED PROCESS

In addition to the six best practices, there are three important features of the Rational Unified Process worth mentioning now:

- The role of use cases in driving many aspects of the development
- Its use as a process framework that can be tailored and extended by an adopting organization
- The need for software development tools to support the process

Use-Case-Driven Development

It is often difficult look at a traditional object-oriented system model and tell how the system does what it is supposed to do. We believe that this difficulty stems from the lack of a consistent, visible thread through the system when it performs certain tasks. In the Rational Unified Process, *use cases* provide that thread by defining the behavior performed by a system.

Use cases are not required in object orientation, but they provide an important link between system requirements and other development artifacts such as design and tests. Other object-oriented methods provide use-case-like representation but use different names for it, such as scenarios or threads.

The Rational Unified Process is a *use-case-driven* approach, which means that the use cases defined for the system are the foundation for the rest of development process. Use cases play a major role in several of the process workflows, especially requirements, design, test, and management. Use cases are also key to business modeling.

If you are unfamiliar with the concept of use case, Chapter 6, A Use-Case-Driven Process, introduces them in more detail and shows how they are used.

Process Configuration

The Rational Unified Process is general and comprehensive enough to be used "as is" or out of the box by many small-to-medium software development organizations, especially those that do not have a very strong process culture. But it is also a process framework that the adopting organization can modify, adjust, and expand to accommodate the specific needs, characteristics, constraints, and history of its organization, culture, and domain.

A process should not be followed blindly, generating useless work and producing artifacts that are of little added value. Instead, the process must be made as lean as possible while still fulfilling its mission to rapidly produce predictably high-quality software. Complementing the process are the best practices of the adopting organization along with its specific rules and procedures.

The process elements that are likely to be modified, customized, added, or suppressed include artifacts, activities, workers, and workflows as well as guidelines and artifact templates. These fundamental process elements are introduced in Chapter 3, where we describe the process model underlying this framework. Chapter 17 explains how the process is implemented in an adopting organization and sketches the steps involved in its configuration.

Tools Support

To be effective, a process must be supported by adequate tools. The Rational Unified Process is supported by a vast palette of tools that automate steps in many activities. These tools are used to create and maintain the various artifacts—models in particular—of the software engineering process, namely visual modeling, programming, and testing. The tools are invaluable in supporting the bookkeeping associated with the change management and the configuration management that accompanies each iteration.

Chapter 3 introduces the concept of *tool mentors,* which provide tool-related guidance. As we work through the various process workflows in Chapters 7–15, we introduce the tools of choice to support the activities of each workflow.

A BRIEF HISTORY OF THE RATIONAL UNIFIED PROCESS

The Rational Unified Process has matured over the years and reflects the collective experience of the many people and companies that today make up Rational Software's rich heritage. Let us take a quick look at the process's ancestry, as illustrated in Figure 2-4.

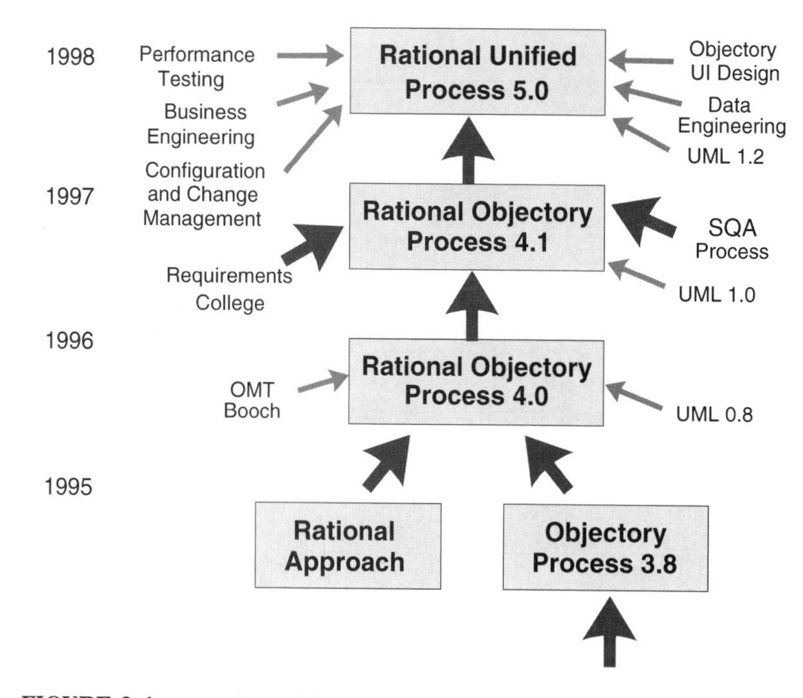

FIGURE 2-4 ***Genealogy of the Rational Unified Process***

Going backward in time, the Rational Unified Process is the direct successor to the Rational Objectory Process (version 4). The Rational Unified Process incorporates more material in the areas of data engineering, business modeling, project management, and configuration management, the latter as a result of a merger with Pure-Atria. It also brings a tighter integration to the Rational Software suite of tools.

The Rational Objectory Process was the result of the integration of the Rational Approach and the Objectory process (version 3.8) after the merger of Rational Software Corporation and Objectory AB in 1995. From its Objectory ancestry, the process has inherited its process model (described in Chapter 3) and the central concept of use case. From its Rational background, it gained the current formulation of iterative development and architecture. This version also incorporated material on requirements management from Requisite, Inc., and a detailed test process inherited from SQA, Inc., companies that also merged with Rational Software. Finally, this process was the first one to use the newly created Unified Modeling Language (UML 0.8).

The Objectory process was created in Sweden in 1987 by Ivar Jacobson as the result of his experience with Ericsson. This process became a product at his company, Objectory AB. Centered on the concept of use case and an object-oriented design method, it rapidly gained recognition in the software industry and has been adopted and integrated by many companies worldwide. A simplified version of the Objectory process was published as a textbook in 1992.[3]

The Rational Unified Process is a specific and detailed instance of a more generic process described by Ivar Jacobson, Grady Booch, and James Rumbaugh in the textbook *The Unified Software Development Process*.[4]

---

3. Ivar Jacobson et al., *Object-Oriented Software Engineering–A Use Case-Driven Approach*. Reading, MA: Addison-Wesley, 1992.

4. Ivar Jacobson, Grady Booch, and James Rumbaugh, *The Unified Software Development Process*. Reading, MA: Addison Wesley Longman, 1998.

SUMMARY

- The Rational Unified Process is a software development process covering the entire software development life cycle.
- It is a process product that brings a wealth of knowledge, always up-to-date, to the developer's workstation.
- It embeds guidance on many modern techniques and approaches: object technology and component-based development, modeling and UML, architecture, iterative development, and so on.
- It is not a frozen product; rather, it is alive, constantly maintained, and continuously evolving.
- It is based on a solid process architecture, and that allows a development organization to configure and tailor it to suit its needs.
- It supports our six best practices for software development.
  1. Develop software iteratively.
  2. Manage requirements.
  3. Use component-based architectures.
  4. Visually model software.
  5. Verify software quality.
  6. Control changes to software.
- It is supported by an extensive palette of tools.

Chapter 3

Static Structure: Process Description

THIS CHAPTER DESCRIBES how the Rational Unified Process is represented. We introduce the key concepts of worker, activities, artifacts, and workflow as well as other elements used in the process's description.

A MODEL OF THE RATIONAL UNIFIED PROCESS

A process describes *who* is doing *what*, *how*, and *when*. The Rational Unified Process is represented using four primary modeling elements:

- Workers: the *who*
- Activities: the *how*
- Artifacts: the *what*
- Workflows: the *when*

These elements are shown in Figure 3-1.

The term *worker* defines the behavior and responsibilities of an individual or a group of individuals working together as a team. The behavior is expressed in terms of *activities* the worker performs, and each worker is associated with a set of cohesive activities. "Cohesive" in this sense means those activities best performed

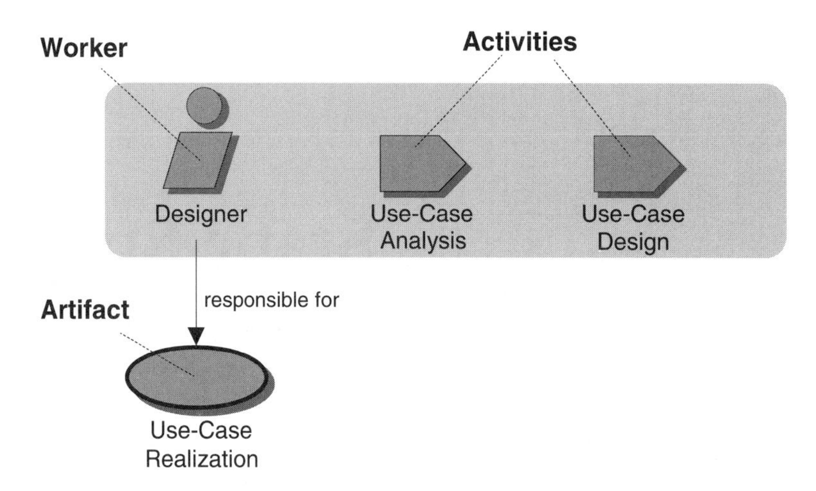

FIGURE 3-1 ***Worker, activities, and artifacts***

by one person. The responsibilities of each worker are usually expressed in relation to certain *artifacts* that the worker creates, modifies, or controls.

WORKERS

It's helpful to think of a worker as a "hat" that an individual can wear during the project. One person may wear many different hats. This distinction is important because it is natural to think of a worker as the individual or the team, but in the Rational Unified Process the term *worker* refers to the roles that define how the individuals should do the work. A worker performs one or more roles and is the owner of a set of artifacts. Another way to think of a worker is as a part in a play—a part that can be performed by many different actors.

The following are examples of workers.

- *System analyst*
  An individual acting as a system analyst leads and coordinates requirements elicitation and use-case modeling by outlining the system's functionality and delimiting the system.

- *Designer*
  An individual acting as a designer defines the responsibilities, operations, attributes, and relationships of one or more classes and determines how they should be adjusted to the implementation environment.
- *Test designer*
  An individual acting as a test designer is responsible for the planning, design, implementation, and evaluation of tests, including generating the test plan and test model, implementing the test procedures, and evaluating test coverage, results, and effectiveness.

Note that workers are not individuals; instead, they describe how individuals should behave in the business and the responsibilities of an individual. Individual members of the software development organization wear different hats, or play different parts or roles.[1] The mapping from individual to worker is performed by the project manager when planning and staffing the project. This mapping allows an individual to act as several different workers and for a worker to be played by several individuals.

In the example shown in Figure 3-2, one individual, Sylvia, can be a Worker: Use-Case Designer in the morning and act as a Worker: Design Reviewer in the afternoon. Paul and Mary are both Designers, although they are likely responsible for different classes or different design packages.

For each worker, a set of expected skills must be provided by the individual who is designated as the worker. Sylvia must understand how to design a use case and how to review a part of the design.

Workers are usually denoted in the process prefixed with the word *Worker,* as in Worker: Integration Tester. Appendix A lists all workers defined in the Rational Unified Process.

1. However, we often write, "The designer of class X does this" when, strictly speaking, we should write, "The individual acting as the designer for class X does this."

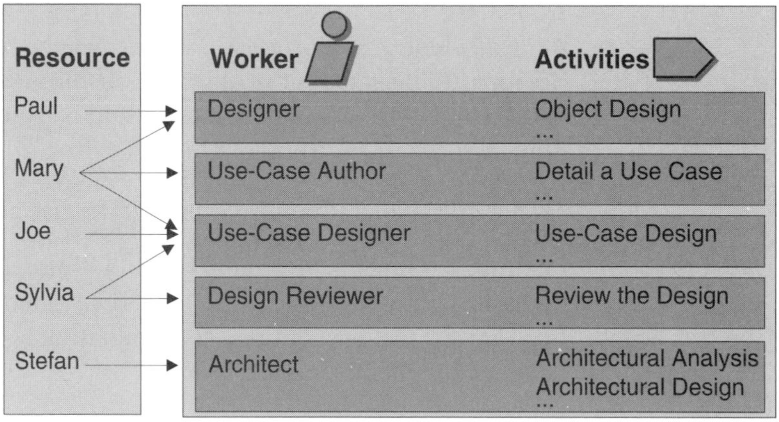

FIGURE 3-2 *People and workers*

ACTIVITIES

An *4activity* of a specific worker is a unit of work that an individual in that role may be asked to perform. The activity has a clear purpose, usually expressed in terms of creating or updating artifacts, such as a model, a class, or a plan. Every activity is assigned to a specific worker.

The granularity of an activity is generally a few hours to a few days. It usually involves one worker and affects one or only a small number of artifacts. An activity should be usable as an element of planning and progress; if it is too small, it will be neglected, and if it is too large, progress will have to be expressed in terms of an activity's parts.

Activities may be repeated several times on the same artifact, especially from one iteration to another as the system is refined and expanded. Repeated activities may be performed by the same worker but not necessarily the same individual.

In object-oriented terms, the worker is an active object, and the activities that the worker performs are operations performed by that object. The following are examples of activities.

- *Plan an iteration:* performed by the Worker: Project Manager
- *Find use cases and actors:* performed by the Worker: System Analyst

- *Review the design:* performed by the Worker: Design Reviewer
- *Execute a performance test:* performed by the Worker: Performance Tester

Activities are usually prefixed with the word *Activity,* as in *Activity: Find use case and actors.* Chapters 7 through 15 give an overview of all activities in the Rational Unified Process.

Activity Steps

Activities are broken into steps. Steps fall into three main categories.

- *Thinking steps*
  The worker understands the nature of the task, gathers and examines the input artifacts, and formulates the outcome.
- *Performing steps*
  The worker creates or updates some artifacts.
- *Reviewing steps*
  The worker inspects the results against some criteria.

Not all steps are necessarily performed each time an activity is invoked, so they can be expressed in the form of alternative flows.

For example, the Activity: Find use cases and actors decomposes into these steps.

1. Find actors.
2. Find use cases.
3. Describe how actors and use cases interact.
4. Package use cases and actors.
5. Present the use-case model in use-case diagrams.
6. Develop a survey of the use-case model.
7. Evaluate your results.

The finding part (steps 1 to 3) requires some thinking; the performing part (steps 4 to 6) involves capturing the result in the use-case model; the reviewing part (step 7) requires the worker to evaluate the result to assess completeness, robustness, intelligibility, or other qualities.

ARTIFACTS

An *artifact* is a piece of information that is produced, modified, or used by a process. Artifacts are the tangible products of the project: the things the project produces or uses while working toward the final product. Artifacts are used as input by workers to perform an activity and are the result or output of such activities. In object-oriented design terms, just as activities are operations on an active object (the worker), artifacts are the parameters of these activities.

Artifacts may take various shapes or forms:

- A model, such as the use-case model or the design model
- A model element—an element within a model—such as a class, a use case, or a subsystem
- A document, such as a business case or software architecture document
- Source code
- Executables

Note that *artifact* is the term used in the Rational Unified Process. Other processes use terms such as work product, work unit, and so on, to denote the same thing. Deliverables are only the subset of all artifacts that end up in the hands of the customers and end users.

Artifacts can also be composed of other artifacts. For example, the design model contains many classes; the software development plan contains several other plans: a staffing plan, a phase plan, a metrics plan, iteration plans, and so on.

Artifacts are most likely to be subject to version control and configuration management. Sometimes, this can be achieved only by versioning the container artifact when it is not possible to do it for the elementary, contained artifacts. For example, you may control the versions of a whole design model or design package and not of the individual classes they contain.

Typically, artifacts are *not* documents. Many processes place an excessive focus on documents and in particular on paper documents. The Rational Unified Process discourages the systematic production of paper documents. The most efficient and pragmatic approach to managing project artifacts is to maintain the artifacts *within* the appropriate tool used to create and manage them. When necessary, you can generate documents (snapshots) from these tools on a just-in-time basis.

You should also consider delivering artifacts to the interested parties inside and together with the tool rather than on paper. This approach ensures that the information is always up-to-date and is based on actual project work, and producing it shouldn't require additional effort.

The following are examples of artifacts:

- A design model stored in Rational Rose
- A project plan stored in Microsoft Project
- A defect stored in ClearQuest
- A project requirements database in Requisite Pro

However, certain artifacts must be plain text documents, as in the case of external input to the project or when it is simply the best means of presenting descriptive information.

Reports

Models and model elements may have associated reports. A *report* extracts information about models and model elements from a tool. For example, a report presents an artifact or a set of artifacts for a review. Unlike regular artifacts, reports are not subject to version control. You can reproduce them at any time by going back to the artifacts that generated them.

Artifacts are usually denoted in the process prefixed with the word *Artifact,* as in Artifact: Use-case storyboard.

Sets of Artifacts

The artifacts of the Rational Unified Process fall into one of five categories, or *information sets*:

- Management set
- Requirements set
- Design set
- Implementation set
- Deployment set

The *management set* groups all artifacts related to the software business and to the management of the project:

- Planning artifacts, such as the software development plan (SDP), the business case, the actual process instance used by the project (the development case), and so on
- Operational artifacts, such as a release description, status assessments, deployment documents, and defect data

The *requirements set* groups all artifacts related to the definition of the software system to be developed:

- The vision document
- Requirements in the form of stakeholders' needs, use-case model, and supplementary specification
- The business model, if it is required for an understanding of the business processes supported by the system

The *design set* contains description of the system to be built (or as built) in the form of

- The design model
- The architecture description
- The test model

The *implementation set* includes

- The source code and executables
- The associated data files or the files needed to produce them

The *deployment set* contains all the information delivered, including

- Installation scripts
- User documentation
- Training material

In an iterative development process, the various artifacts are not produced, completed, or even frozen in one phase before you move on to the next phase. On the contrary, the five information sets evolve throughout the development cycle. They are *grown*, as

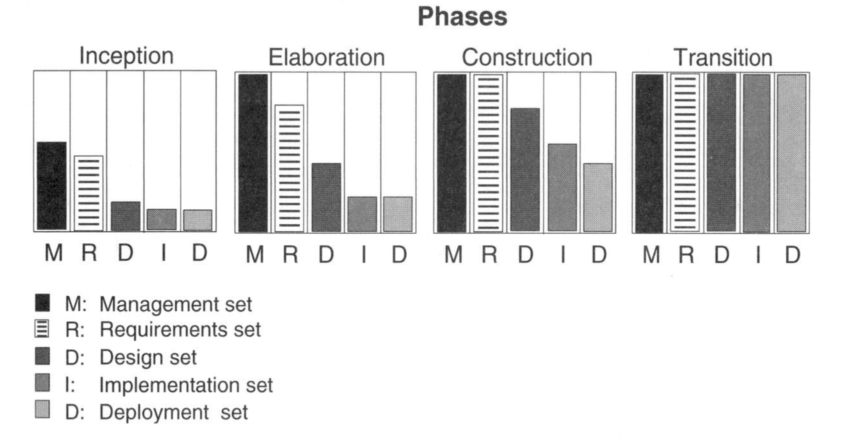

FIGURE 3-3 *Growing the information sets*

depicted in Figure 3-3. Appendix B lists all artifacts defined in the Rational Unified Process.

WORKFLOWS

A mere enumeration of all workers, activities, and artifacts does not quite constitute a process. We need a way to describe meaningful sequences of activities that produce some valuable result and to show interactions between workers.

A *workflow* is a sequence of activities that produces a result of observable value. In UML terms, a workflow can be expressed as a sequence diagram, a collaboration diagram, or an activity diagram. We use a form of activity diagrams in this book. Figure 3-4 is an example of a workflow.[2]

Note that it is not always possible or practical to represent all the dependencies between activities. Often, two activities are more tightly interwoven than shown, especially when they involve the same worker or the same individual. People are not machines, and the workflow cannot be interpreted literally as a program that people are to follow exactly and mechanically.

2. Strictly speaking, our core workflows are *workflow classes,* of which there are many possible *workflow instances.*

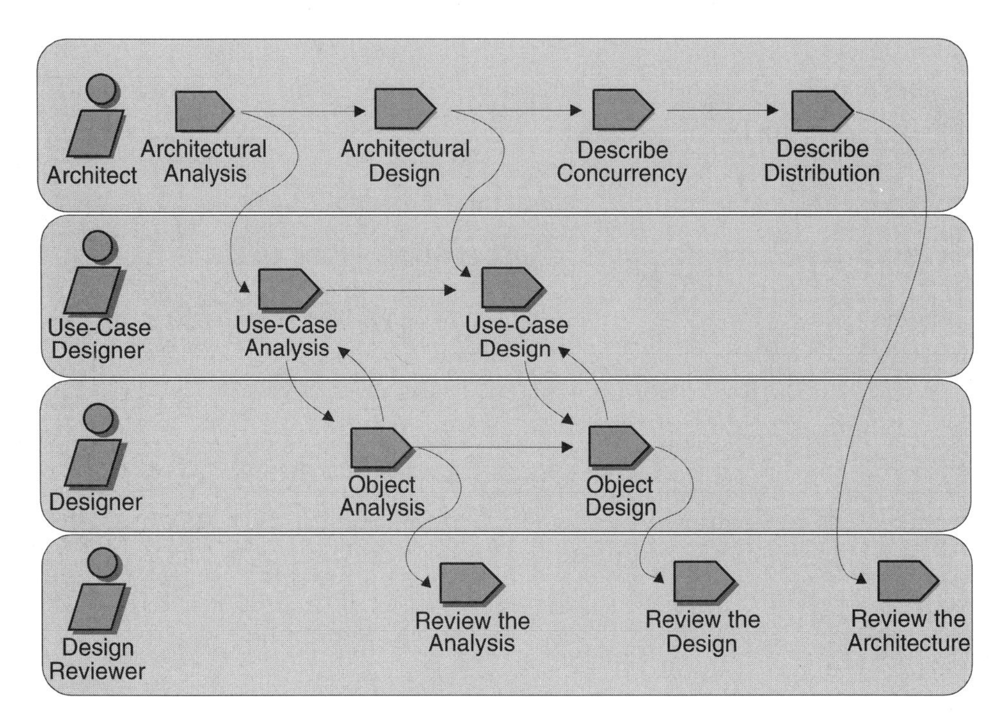

FIGURE 3-4 *Example of a workflow*

There are many ways to organize the set of activities in workflows. We have organized the Rational Unified Process using

- Core workflows
- Iteration workflows
- Workflow details

Core Workflows

There are nine *core process workflows* in the Rational Unified Process, and they represent a partitioning of all workers and activities into logical groupings (see Figure 3-5). The core process workflows are divided into six core engineering workflows and three core supporting workflows. The engineering workflows are as follows:

1. Business modeling workflow
2. Requirements workflow

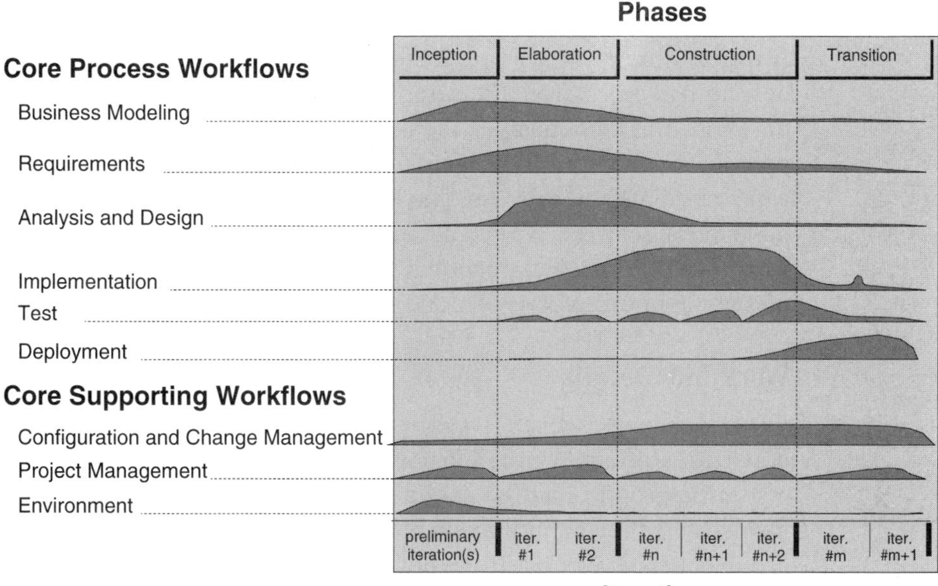

FIGURE 3-5 *Nine core process workflows*

3. Analysis and design workflow
4. Implementation workflow
5. Test workflow
6. Deployment workflow

The three core supporting workflows are as follows:

1. Project management workflow
2. Configuration and change management workflow
3. Environment workflow

Although the names of the six core engineering workflows may evoke the sequential phases in a traditional waterfall process, you will see in Chapter 4 that the phases of an iterative process are different and that these workflows are revisited again and again throughout the life cycle. The actual complete workflow of a project interleaves these nine core workflows and repeats them with various emphases and levels of intensity at each iteration. The core workflows are discussed in detail in Chapters 7 through 15.

Iteration Workflows

Iteration workflows are another means of presenting the process, describing it more from the perspective of what happens in a typical iteration. You can consider them to constitute instantiations of the process for one iteration. There are an infinite number of ways you can instantiate the process. The Rational Unified Process contains descriptions of a few typical iteration workflows. They are given primarily for pedagogical purposes, as you can see with the few examples given in Chapter 16.

Workflow Details

Each of the core workflows covers a lot of ground. To break them down, the Rational Unified Process uses *workflow details* to express a specific group of activities that are closely related; that are performed together or in a cyclical fashion; that are performed by a group of people working together in a workshop; or that produce an interesting intermediate result. Workflow details also show information flows—the artifacts that are input to and output from the activities—showing how activities interact through the various artifacts.

ADDITIONAL PROCESS ELEMENTS

Workers, activities (organized in workflows), and artifacts represent the backbone of the Rational Unified Process static structure. But other elements are added to activities or artifacts to make the process easier to understand and use and to provide more comprehensive guidance to the practitioner. These additional process elements are

- Guidelines
- Templates
- Tool mentors
- Concepts

These elements enhance the primary elements, as shown in Figure 3-6.

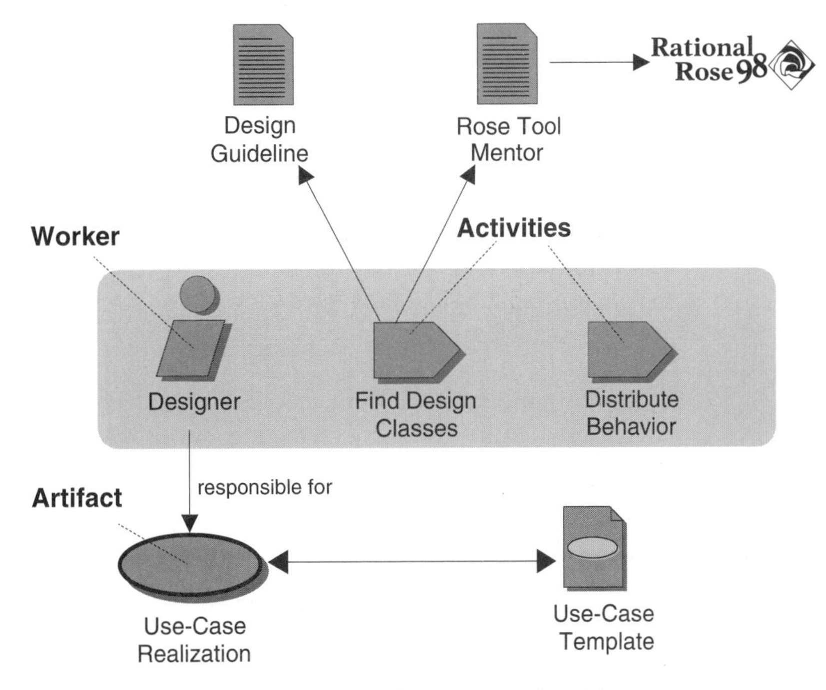

FIGURE 3-6 *Adding templates, tool mentors, and guidelines*

Guidelines

Activities and steps are kept brief and to the point because they are intended to serve as references for what needs to be done. Therefore, they must be useful for neophytes looking for guidance as well as for experienced practitioners needing a reminder.

Attached to activities, steps, or artifacts are *guidelines*. Guidelines are rules, recommendations, or heuristics that support activities and steps. They describe well-formed artifacts, focusing on qualities. Guidelines also describe specific techniques, such as transformations from one artifact to another or to the use of the Unified Modeling Process. Guidelines are used also to assess the quality of artifacts—in the form of checklists associated with artifacts—or to review activities. The following are examples of guidelines:

- Modeling guidelines, which describe well-formed modeling elements such as use cases, classes, and test cases

- Programming guidelines, for languages such as C++ or Ada, describing well-formed programs
- User-interface guidelines
- Guidelines on how to create a specific artifact, such as a risk list or an iteration plan
- Work guidelines, which give practical advice on how to undertake an activity, especially for groups
- Checklists to be used as part of a review or by a worker to verify that an activity is complete

Some guidelines may need to be refined or specialized for a given organization or project to accommodate project specifics, such as the use of a particular technique or tool.

The following are examples of this latter type of guidelines:

- User-interface guidelines, such as a description of the windowing style specific to a project: color palette, fonts, gallery of icons, and so on
- Programming guidelines, such as a description of naming conventions specific to the project

Templates

Templates are "models," or prototypes, of artifacts. Associated with the artifact description are one or more templates that can be used to create the corresponding artifacts. Templates are linked to the tool that is to be used. For example:

- Microsoft Word templates for documents and some reports
- SoDA templates for Microsoft Word or FrameMaker that extract information from tools such as Rational Rose, Requisite Pro, or TeamTest
- Microsoft FrontPage templates for the various elements of the process
- Microsoft Project template for the project plan

As with guidelines, organizations may want to customize the templates before using them by adding the company logo, some project identification, or information specific to the type of project.

Tool Mentors

Activities, steps, and associated guidelines provide general guidance to the practitioner. To go one step further, *tool mentors* are an additional means of providing guidance by showing you how to perform the steps using a specific software tool. Tool mentors are provided in the Rational Unified Process, linking its activities with tools such as Rational Rose, Requisite Pro, ClearCase, ClearQuest, and TestStudio. The tool mentors almost completely encapsulate in the tool set the dependencies of the process, keeping the activities free from tool details. An organization can extend the concept of tool mentor to provide guidance for other tools.

Concepts

Some of the key concepts, such as iteration, phase, risk, performance testing, and so on, are introduced in separate sections of the process, usually attached to the most appropriate core workflow. Several concepts are also introduced in this book.

A PROCESS FRAMEWORK

With this structure, the Rational Unified Process constitutes a process *framework*. Workers, artifacts, activities, guidelines, concepts, and mentors are the elements that you can add or replace to evolve or adapt the process to the organization's needs. How to do this is further developed in Chapter 17 and is described in the environment workflow.

SUMMARY

- The Rational Unified Process model is built on three fundamental entities: workers, activities, and artifacts.
- Workflows relate activities and workers in sequences that produce valuable results.
- Guidelines, templates, and tool mentors complement the description of the process by providing detailed guidance to the practitioner.
- The Rational Unified Process is a process framework organized to enable the configuration of its static structure.

Chapter 4

Dynamic Structure: Iterative Development

THIS CHAPTER DESCRIBES the life-cycle structure of the Rational Unified Process—that is, how the process rolls out over time. We introduce the concept of iterative development, with its phases, milestones, and iterations, as well as the factors that drive the process: risk mitigation and incremental evolution.

THE SEQUENTIAL PROCESS

It has become fashionable to blame many problems and failures in software development on the sequential (or "waterfall") process depicted in Figure 4-1. This is rather surprising because at first this method seems like a reasonable approach to system development.

A Reasonable Approach

To examine how many engineering problems are solved using the sequential process, let's look at the steps it typically follows.

1. Completely understand the problem to be solved, its requirements, and its constraints. Capture them in writing and get all interested parties to agree that this is what they need to achieve.
2. Design a solution that satisfies all requirements and constraints. Examine this design carefully and make sure that all interested parties agree that it is the right solution.

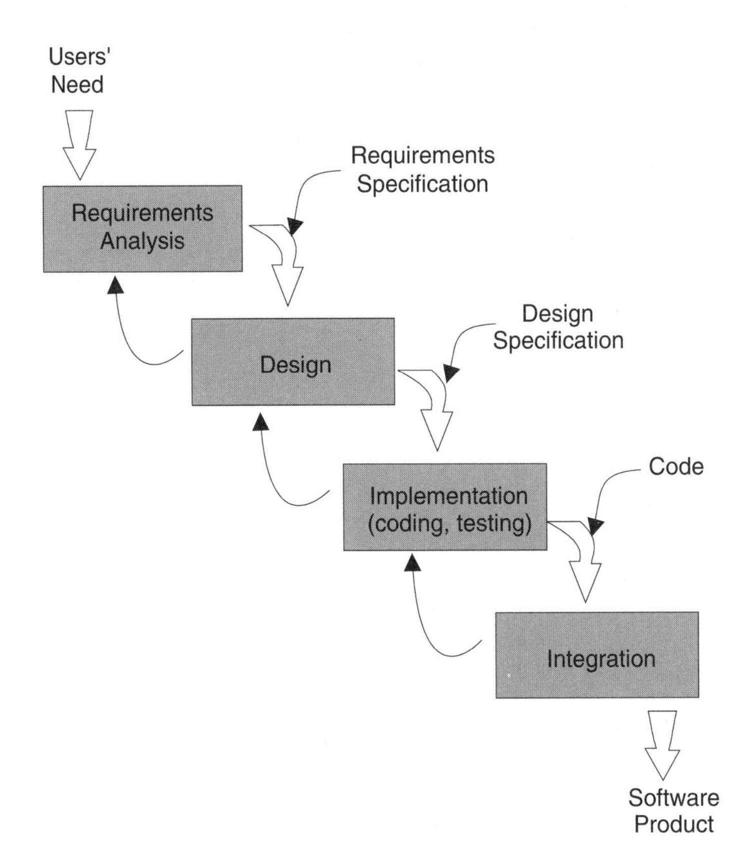

FIGURE 4-1 *The sequential process*

3. Implement the solution using your best engineering techniques.

4. Verify that the implementation satisfies the stated requirements.

5. Deliver. Problem solved!

That is how skyscrapers and bridges are built. It's a rational way to proceed but only because the problem domain is relatively well known; engineers can draw on hundreds of years of experimentation in design and construction.

By contrast, software engineers have had only a few decades to explore their field. Software developers worked very hard, particularly in the seventies and eighties, to accumulate experimental results in software design and construction. In 1980, I would have sworn that the sequential process was the one and only reasonable approach.

If the sequential process is ideal, however, why aren't the projects that use it more successful? There are many reasons.

- We made the wrong assumptions.
- The context of software development is somewhat different from that of other engineering disciplines.
- We have failed to incorporate some human factors.
- We have tried to stretch an approach that works in certain well-defined circumstances beyond what it can bear.
- We are still only in the exploratory phase of software engineering. We do not have the experience of hundreds of years of trial and error that makes building a bridge appear to be a mechanical process. This is the primary reason.

Let us review two fundamentally wrong assumptions that often hinder the success of software projects.

Wrong Assumption 1: Requirements Will Be Frozen

Notice that in the preceding description we assume in step 1 that we can capture the entire problem at the beginning. We assume that we can nail down all the requirements in writing in an unambiguous fashion and begin the project with a stable foundation. Despite all our efforts, though, this almost always proves to be impossible. Requirements will change. We must accept this fact. Unless we are solving a trivial problem, new or different requirements will appear. Requirements change for many different reasons. Let's look at a few of them.

- *The users will change.*
  The users' needs cannot be frozen in time. This is especially true when the development time is measured not in weeks or months but in years. Users will see other systems and other products, and they'll want some of the features they see. Their own work environment will evolve, and they will become better educated.
- *The problem will change.*
  After the system is implemented or while it is being implemented, the system itself will affect the perspective of users.

It is quite a different thing to try out features or see them demonstrated than to read about them. As soon as the end users see how their intentions have been translated into a system, the requirements will change. In fact, the one point when users know exactly what they want is not two years before the system is ready but rather a few weeks or months *after* delivery of the system when they are beyond the initial learning phase. This is known as the IKIWISI effect: "I'll Know It When I See It."[1]

Users don't really know what they want, but they know what they do not want when they see it. Therefore, efforts to detail, capture, and freeze the requirements may ultimately lead to the delivery of a perfect system with respect to the requirements but the wrong system with respect to the real problem at the time of delivery.

- *The underlying technology will change.*
  New software or hardware techniques and products will emerge that you will want to exploit. On a multiyear project, the hardware platform bid at the beginning of the project may no longer be manufactured at delivery time.
- *The market will change.*
  The competition will introduce better products to the market. What is the point of developing the perfect product relative to the original spec if you end up with the wrong product relative to what the marketplace expects when you are finally finished?
- *We cannot capture requirements with sufficient detail and precision.*
  Formal methods have held the promise of a solution, but on the eve of the third millenium they have not gained significant acceptance in the industry except in small, specialized domains. They are hard to apply and very user-unfriendly. Try teaching temporal logic or colored Petri nets to an audience of bank tellers and branch managers so that they can read and approve the specification of their new system.

1. Origin uncertain; Barry Boehm used this acronym at a workshop on software architecture at the University of Southern California in 1997.

Wrong Assumption 2: We Can Get the Design Right on Paper Before Proceeding

The second step of the sequential process assumes that we can confirm that our design is the right solution to the problem. By "right" we imply all the obvious qualities: correctness, efficiency, feasibility, and so on. With complete requirements tracing, formal derivation methods, automated proof, generator techniques, and design simulation, some of these qualities can be achieved. However, few of these techniques are readily available to practitioners, and many of them require that you begin with a formal definition of the problem. You can accumulate pages and pages of design documentation and hundreds of blueprints and spend weeks in reviews, only to discover, late in the process, that the design has major flaws that cause serious breakdowns.

Software engineering has not reached the level of other engineering disciplines (and perhaps it never will) because the underlying "theories" are weak and poorly understood, and the heuristics are crude. Software engineering may be misnamed. At various times it more closely resembles a branch of psychology, philosophy, or art than engineering. Relatively straightforward laws of physics underlie the design of a bridge, but there is no strict equivalent in software design. Software is "soft" in this respect.

Bringing Risks into the Picture

The sequential (or waterfall) process does work. It has worked fine for me on small projects ranging from a few weeks to a few months, on projects in which we could clearly anticipate what would happen, and on projects in which all hard aspects were well understood. For projects having little or no novelty, you can develop a plan and execute it with little or no surprise. If the current project is somewhat like the one you completed last year—and the one the year before—and if you use the same people, the same tools, and the same design, the sequential approach will work well.

The sequential process breaks down when you tackle projects having a significant level of novelty, unknowns, and risks. You cannot anticipate the difficulties you may encounter, let alone how you will counter them. The only thing you can do is to build some slack into the schedule and cross your fingers.

The absence of fundamental "laws of software" and the pace at which software evolves make it a risky domain. Techniques for reinforcing concrete have not dramatically changed since my grandfather used them in the early twenties in an engineering bureau. Software tools, techniques, and products, on the other hand, have a lifetime of a few years at best. So every time we try to build a system that is a bit more complicated, somewhat larger, or a little more challenging, we are in dangerous and risky territory, and we must take this into account.

That's why we bring risk analysis into the picture.

Stretching the Time Scale

If you stretch what works for a three-month project to fit a three-year project, you expose the project not only to the changing contexts we have discussed but also to other subtle effects related to the people involved. Software developers who know that they will see tangible results within the next two to three months can remain well focused on the real outcome. Very quickly, they will get feedback on the quality of their work. If small mistakes are discovered along the way, the developers won't have to go very far back in time to correct them.

But picture developers in the middle of the design phase of a three-year project. The target is to finish the design within four months. The developers may not even be around to see the final product up and running. Progress is measured in pages or diagrams and not in operational features. There is nothing tangible, nothing to get the adrenaline flowing.

There is little feedback on the quality of the current activity, because defects will be found later, during integration or test, perhaps 18 months from now. The developers have few opportunities to improve the way they work. Moreover, strange things discovered in the requirements text mean that developers must revisit discussions and decisions made months ago. Is it any wonder that they have a hard time staying motivated? The original protagonists are no longer in the project, and the contract with the customer is as hard and inflexible as a rock.

The developers have only one shot at each kind of activity, with little opportunity to learn from their mistakes. You have one shot at design, and it had better be good. You say you've never designed a system like this? Too bad! You have one shot at coding, and it had better be good. You say this is a new programming language? Well, you can work longer hours to learn its new features. There's only one shot at testing, and it had better be a no-fault run. You say this is a new system and no one really knows how it's supposed to work? Well, you'll figure it out. If the project introduces any new techniques or tools or any new people, the sequential process gives you no latitude for learning and improvement.

Pushing Paperwork on the Shelves

In the sequential process, the goal of each step except the last one is to produce and complete an intermediate artifact (usually a paper document) that is reviewed, approved, frozen, and then used as the starting point for the next step. In practice, sequential processes place an excessive emphasis on the production and freezing of documents. Some limited amount of feedback to the preceding step is tolerated, but feedback on the results of earlier steps is seen as disruptive. This is related to the reluctance to change requirements and to the loss of focus on the final product that is often seen during long projects.

Volume-Based Versus Time-Based Scheduling

Often, timeliness is the most important factor in the success of a software project. In many industries, delivery of a product on time and with a short turnaround for new features is far more important than delivery of a complete, full-featured, perfect system. To achieve timeliness, you must be able to adjust the contents dynamically by dropping or postponing some features to deliver incremental value on time. With a linear approach, you do not gain much on the overall schedule if you decide in the middle of the implementation to drop feature X. You have already expended the time and effort to specify, design, and code the feature. That's why this model isn't suitable when a company wants to work with schedules that are time-based (for example, in three months we can do the first three items on your list, and three months later we'll have the next two, and so on) and not volume-based (it will take us nine months to do everything that you want).

For these reasons and a few others that we will cover later, software organizations have tried another approach.

OVERCOMING DIFFICULTIES: ITERATE!

How do you eat an elephant? One bite at a time! If the sequential or waterfall approach is reasonable and even successful for short projects or those with a small amount of novelty or risk, why not break down the life cycle of a large project into a succession of small waterfall projects? In this way, you can address some requirements and some risks, design a little, implement a little, validate it, and then take on more requirements, design some more, build some more, validate, and so on, until you are finished. This is the *iterative* approach.

Figure 4-2 compares the iterative approach with the sequential approach. It shows what things look like in an iterative process for one development cycle—from the initial idea to the point when a complete, stable, quality product is delivered to the end users.

This technique is easy to illustrate but not very easy to achieve. It raises more questions than it answers.

- How does this work converge to become a product? How do you avoid having each iteration start over from scratch?
- How do you select what to do in each iteration? Which requirements do you consider, and which risks do you address?

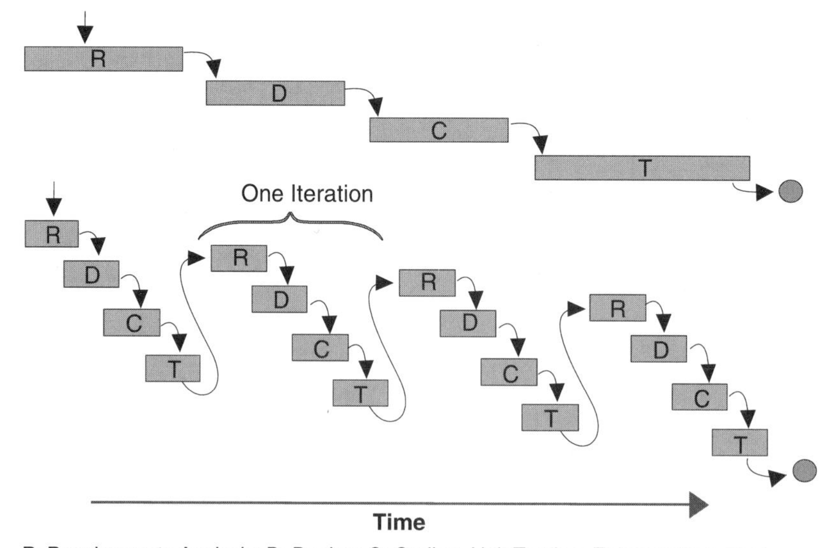

FIGURE 4-2 *From a sequential to an iterative life cycle*

- How does this approach solve the major issues we identified earlier?

The Rational Unified Process answers these questions. We'll discuss the answers in the rest of this chapter and further in Chapter 7.

GAINING CONTROL: PHASES AND MILESTONES

From a project management perspective, we need a way of assessing progress so that we can ensure that we are not wandering aimlessly from iteration to iteration but are actually converging toward a product. From a management perspective, we must also define points in time to operate as gating functions based on clear criteria. These *milestones* provide points at which we can decide to proceed, abort, or change course. Finally, we must partition and organize the sequence of iterations according to specific short-term objectives. Progress will be measured in the number of use cases completed, features completed, test cases passed, performance requirements satisfied, and risks eliminated.

The iterative process is organized in phases, as shown in Figure 4-3. But unlike the steps in the waterfall approach, the phases here are not the traditional sequence of requirements analysis, design, coding, integration, and test. They are completely orthogonal to the traditional phases. Each phase is concluded by a major milestone.[2]

Let's look at the four phases in more detail.

- *Inception*
  The good idea—specifying the end-product vision and its business case and defining the scope of the project.[3] The inception phase is concluded by the life-cycle objective (LCO) milestone.
- *Elaboration*
  Planning the necessary activities and required resources; specifying the features and designing the architecture.[4] The

2. The milestone definition and names are identical to the ones proposed by Barry Boehm in the article "Anchoring the Software Process," *IEEE Software*, July 1996, pp. 73–82.

3. *American Heritage Dictionary* defines *inception* as "the beginning of something, such as an undertaking, a commencement."

4. *American Heritage Dictionary* defines *elaboration* as the process "to develop thoroughly, to express at greater length or greater detail."

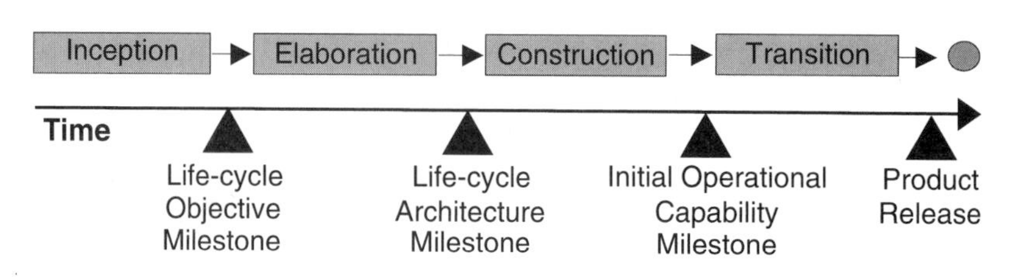

FIGURE 4-3 *The four phases and milestones of the iterative process*

elaboration phase is concluded by the life-cycle architecture (LCA) milestone.

- *Construction*
  Building the product and evolving the vision, the architecture, and the plans until the product—the completed vision—is ready for delivery to its user community. The construction phase is concluded by the initial operational capability (IOC) milestone.
- *Transition*
  Transitioning the product to its users, which includes manufacturing, delivering, training, supporting, and maintaining the product until users are satisfied. It is concluded by the product release milestone, which also concludes the cycle.

The four phases (I, E, C, and T) constitute a development *cycle* and produce a software *generation*. A software product is created in an *initial development cycle*. Unless the life of the product stops at this point, an existing product will evolve into its next generation by a repetition of the same sequence of inception, elaboration, construction, and transition phases, but with a different emphasis on the various phases. We call these periods *evolution cycles* (see Figure 4-4).

As the product goes through several evolution cycles, new generations of the product are produced. Evolution cycles can be triggered by user-suggested enhancements, changes in the user's context, changes in the underlying technology, or reaction to the competition. In practice, cycles may overlap slightly: the inception and elaboration phase may begin during the final part of the transition phase of the previous cycle. We revisit this issue in Chapter 7.

Do not interpret the various figures to mean that the phases are of equal duration; their length will vary greatly depending on the

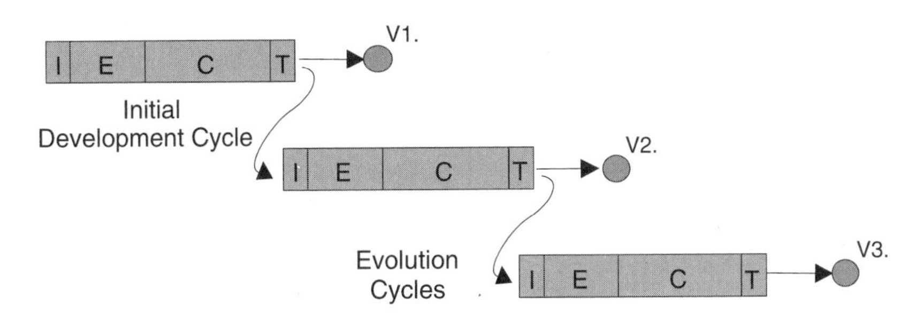

FIGURE 4-4 ***Initial and evolution cycles***

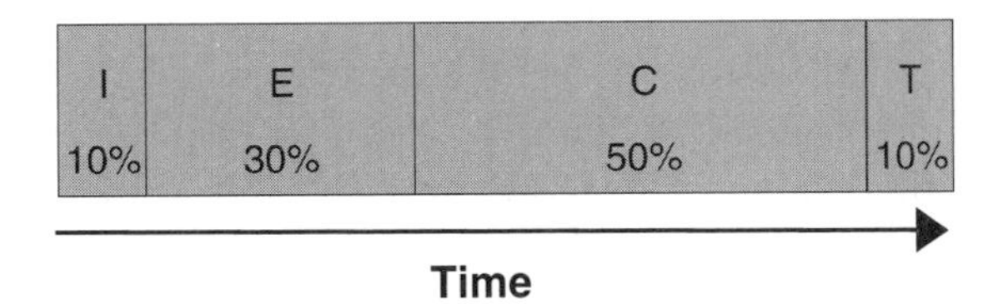

FIGURE 4-5 ***Typical timeline for initial development cycles***

specific circumstances of the project. What is important is the goal of each phase and the milestone that concludes it. Figure 4-5 shows the timeline of a typical project.

For example, a two-year project would have the following:

- A two-and-a-half month inception phase
- A seven-month elaboration phase
- A 12-month construction phase
- A two-and-a-half month transition phase

How does this timeline relate to iterations? In each phase you progress iteratively, and each phase consists of one or several iterations.[5] The life cycle is as shown in Figure 4-6.[6]

A SHIFTING FOCUS ACROSS THE CYCLE

Remember that *each iteration* follows a pattern similar to the waterfall approach, and therefore its workflow contains the activities of

5. In rare cases, a phase may contain no real iteration.

6. The number of iterations per phase in this diagram is for illustration only. We discuss the number of iterations in Chapter 7.

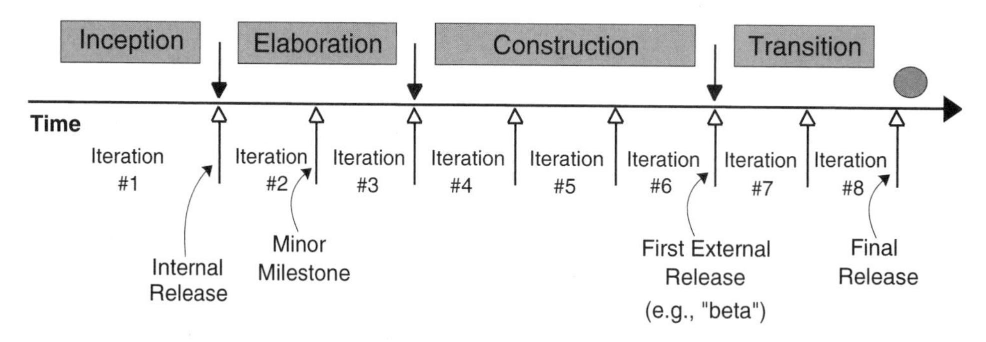

FIGURE 4-6 *Introducing iterations, internal and external releases, and minor milestones*

requirements elicitation and analysis, of design and implementation, and of integration and test. But from one iteration to the next and from one phase to the next, the emphasis on the various activities will change.

Figure 4-7 shows the relative emphasis of the various types of activities over time. If you look at a cross-section of the activities in the middle of each phase, you will notice a number of things. If you

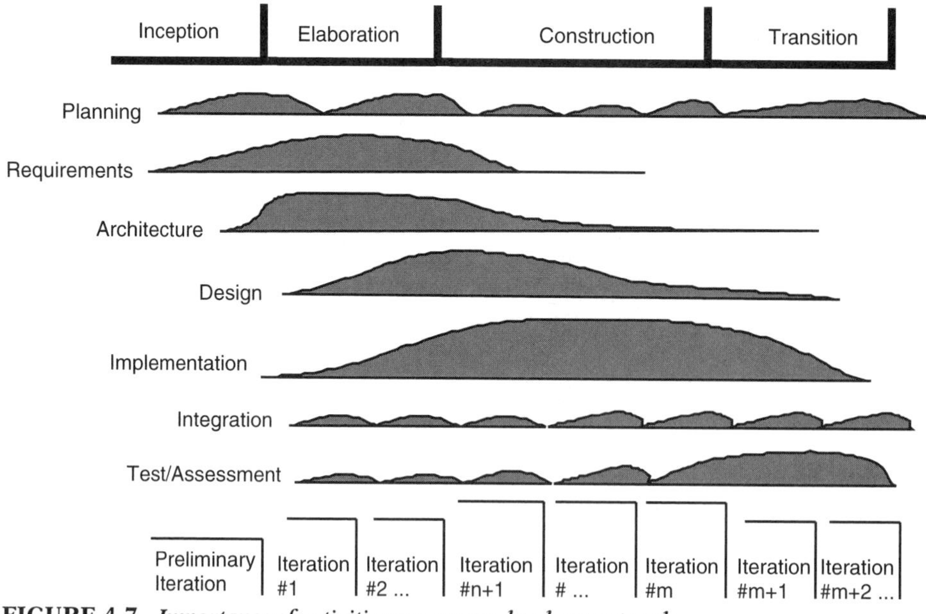

FIGURE 4-7 *Importance of activities across one development cycle*

look at a cross-section of the activities in the middle of each phase, you will notice a number of things.

- In the inception phase, the focus is mainly on understanding the overall requirements and determining the scope of the development effort.
- In the elaboration phase, the focus is primarily on requirements, but some software design and implementation is aimed at prototyping the architecture, mitigating certain technical risks by trying solutions, and learning how to use certain tools and techniques. You finally produce an executable architectural prototype that will serve as the baseline[7] for the next phase.
- In the construction phase, the focus is mostly on design and implementation. Here, you evolve and flesh out the initial prototype into the first operational product.
- In the transition phase, the focus is on ensuring that the system has the right level of quality to meet your objectives; you fix bugs, train users, adjust features, and add missing elements. You produce and deliver the final product.

PHASES REVISITED

This section examines in more detail the purpose of each phase and the evaluation criteria used at each major milestone.[8]

The Inception Phase

Inception | Elaboration | Construction | Transition

The overriding goal of the inception phase is to achieve concurrence among all stakeholders on the life-cycle objectives for the project. The primary objectives of the inception phase include the following:

---

7. A baseline is a reviewed and approved release of artifacts that constitutes an agreed-on basis for further evolution or development and that can be changed only through a formal procedure, such as change and configuration control.

8. The phases were developed in cooperation with Walker Royce, and this section is adapted from Chapter 6 of his book *Software Project Management: A Unified Framework*. Reading, MA: Addison Wesley Longman, 1998.

- Establishing the project's software scope and boundary conditions, including an operational concept, acceptance criteria, and descriptions of what is and is not intended to be in the product
- Discriminating the critical use cases of the system—that is, the primary scenarios of behavior that will drive the system's functionality and will shape the major design trade-offs
- Exhibiting, and perhaps demonstrating, at least one candidate architecture against some of the primary scenarios
- Estimating the overall cost and schedule for the entire project and providing detailed estimates for the elaboration phase that will immediately follow
- Estimating potential risks (the sources of unpredictability)

The essential activities of the inception phase are as follows:

- Formulating the scope of the project—that is, capturing the context and the most important requirements and constraints so that you can derive acceptance criteria for the end product
- Planning and preparing a business case and evaluating alternatives for risk management, staffing, project plan, and trade-offs between cost, schedule, and profitability
- Synthesizing a candidate architecture, evaluating trade-offs in design, and assessing make/buy/reuse decisions so that cost, schedule, and resources can be estimated

The outcome of the inception phase is creation of these artifacts:

- A vision document—that is, a general vision of the core project's requirements, key features, and main constraints
- The use-case model survey, which lists all use cases and actors that can be identified at this early stage
- An initial project glossary
- An initial business case, which includes
  - Business context
  - Success criteria (revenue projection, market recognition, and so on)
  - Financial forecast

- An initial risk assessment
- A project plan, which shows the phases and iterations

For a commercial software product, the business case should include a set of assumptions about the project and the order of magnitude return on investment (ROI) if these assumptions are true. For example, the ROI will be a magnitude of five if the project is completed in one year, two if it is completed in two years, and a negative number after that. These assumptions are checked again at the end of the elaboration phase when the scope and plan are defined with more accuracy.

The resource estimate might encompass either the entire project through to delivery or only the resources needed for the elaboration phase. Estimates of the resources required for the entire project should be viewed as very rough, a "guesstimate" at this point. This estimate is updated during each phase and each iteration and becomes more accurate with each iteration.

The inception phase may also produce the following artifacts:

- An initial use-case model (10%–20% complete) when dealing with an initial development cycle
- A domain model, which is more sophisticated than a glossary (see Chapter 9)
- A business model if necessary (see Chapter 8)
- A preliminary development case description to specify the process used (see Chapter 17)
- One or several prototypes (see Prototypes in Chapter 11)

Milestone: Life-Cycle Objective

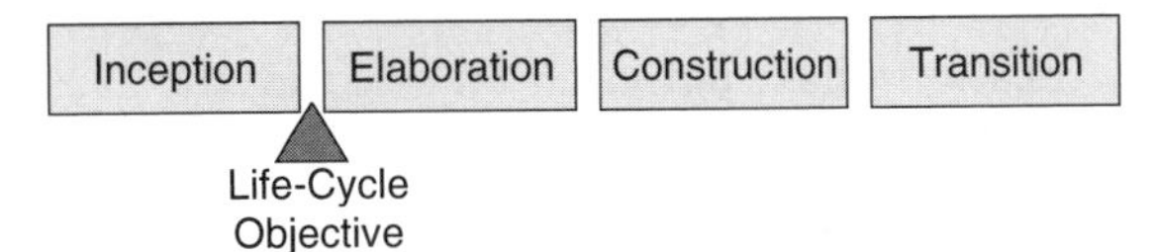

At the end of the inception phase is the first major project milestone: the life-cycle objective milestone. The evaluation criteria for the inception phase are as follows:

- Stakeholder concurrence on scope definition and cost and schedule estimates

- Requirements understanding as evidenced by the fidelity of the primary use cases
- Credibility of the cost and schedule estimates, priorities, risks, and development process
- Depth and breadth of any architectural prototype that was developed
- Actual expenditures versus planned expenditures

If the project fails to pass this milestone, it may be canceled or considerably rethought.

The Elaboration Phase

Inception | Elaboration | Construction | Transition

The purpose of the elaboration phase is to analyze the problem domain, establish a sound architectural foundation, develop the project plan, and eliminate the project's highest-risk elements. To accomplish these objectives, you must have a "mile wide and inch deep" view of the system. Architectural decisions must be made with an understanding of the whole system: its scope, major functionality, and nonfunctional requirements such as performance requirements.

It is easy to argue that the elaboration phase is the most critical of the four phases. At the end of this phase, the hard "engineering" is considered complete and the project undergoes its most important day of reckoning: the decision of whether to commit to the construction and transition phases. For most projects, this phase also corresponds to the transition from a mobile, nimble, low-risk operation to a high-cost, high-risk operation that has substantial inertia. Although the process must always accommodate changes, the elaboration phase activities ensure that the architecture, requirements, and plans are stable enough, and the risks are sufficiently mitigated, that you can predictably determine the cost and schedule for the completion of the development. Conceptually, this level of fidelity corresponds to the level necessary for an organization to commit to a fixed-price construction phase.

In the elaboration phase, an executable architecture prototype is built in one or more iterations, depending on the scope, size, risk, and novelty of the project. At minimum, this effort should address the critical use cases identified in the inception phase, which typically expose the project's major technical risks. Although an evolutionary prototype of a production-quality component is always the goal, this does not exclude the development of one or more exploratory, throw-away prototypes to mitigate specific risks such as trade-offs between design and requirements. Nor does it rule out a component feasibility study or demonstrations to investors, customers, and end users.

The primary objectives of the elaboration phase include the following:

- Defining, validating, and baselining[9] the architecture as rapidly as practical
- Baselining the vision
- Baselining a high-fidelity plan for the construction phase
- Demonstrating that the baseline architecture will support this vision for a reasonable cost in a reasonable time

The essential activities of the elaboration phase are as follows.

- The vision is elaborated, and a solid understanding is established of the most critical use cases that drive the architectural and planning decisions.
- The process, the infrastructure, and the development environment are elaborated, and the process, tools, and automation support are put into place.
- The architecture is elaborated and the components are selected. Potential components are evaluated, and the make/buy/reuse decisions are sufficiently understood to determine the construction phase cost and schedule with confidence. The selected architectural components are integrated and assessed against the primary scenarios. Lessons learned from these activities may result in a redesign of the architecture, taking into consideration alternative designs or reconsideration of the requirements.

9. To *baseline* is to create a baseline: to put a validated release under configuration control so that it can serve as the starting point and reference for further development.

The outcome of the elaboration phase is as follows:

- A use-case model (at least 80% complete) in which all use cases have been identified in the use-case model survey, all actors have been identified, and most use-case descriptions have been developed
- Supplementary requirements capturing the nonfunctional requirements and any requirements that are not associated with a specific use case
- A software architecture description
- An executable architectural prototype
- A revised risk list and a revised business case
- A development plan for the overall project, including the coarse-grained project plan, showing iterations and evaluation criteria for each iteration
- An updated development case specifying the process to be used
- A preliminary user manual (optional)

Milestone: Life-Cycle Architecture

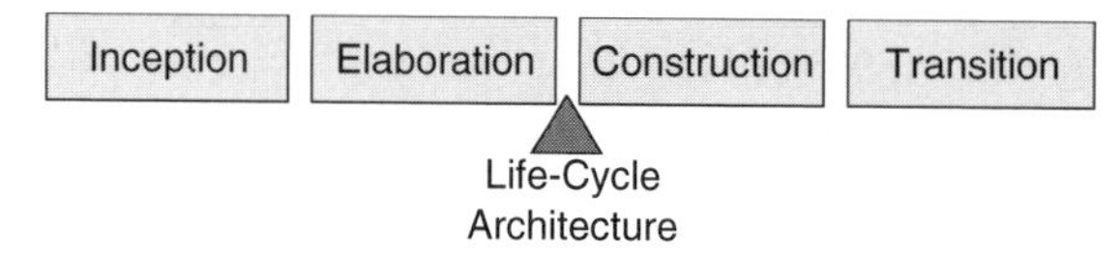

At the end of the elaboration phase is the second important project milestone: the life-cycle architecture milestone. At this point, you examine the detailed system objectives and scope, the choice of architecture, and the resolution of the major risks.

The main evaluation criteria for the elaboration phase involve the answers to the following questions.

- Is the vision of the product stable?
- Is the architecture stable?
- Does the executable demonstration show that the major risk elements have been addressed and credibly resolved?
- Is the construction phase plan sufficiently detailed and accurate? Is it backed up with a credible basis for the estimates?
- Do all stakeholders agree that the current vision can be achieved if the current plan is executed to develop

the complete system, in the context of the current architecture?

- Is the actual resource expenditure versus planned expenditure acceptable?

If the project fails to pass this milestone, it may be aborted or considerably rethought.

The Construction Phase

Inception | Elaboration | **Construction** | Transition

During the construction phase, all remaining components and application features are developed and integrated into the product, and all features are thoroughly tested. The construction phase is, in one sense, a manufacturing process in which emphasis is placed on managing resources and controlling operations to optimize costs, schedules, and quality. In this sense, the management mindset undergoes a transition from the development of intellectual property during inception and elaboration to the development of deployable products during construction and transition.

Many projects are large enough that parallel construction increments can be spawned. These parallel activities can significantly accelerate the availability of deployable releases; they can also increase the complexity of resource management and workflow synchronization. A robust architecture and an understandable plan are highly correlated. In other words, one of the critical qualities of the architecture is its ease of construction. This is one reason that the balanced development of the architecture and the plan is stressed during the elaboration phase.

Primary construction phase objectives include the following:

- Minimizing development costs by optimizing resources and avoiding unnecessary scrap and rework
- Achieving adequate quality as rapidly as practical
- Achieving useful versions (alpha, beta, and other test releases) as rapidly as practical

The essential activities of the construction phase are as follows:

- Resource management, resource control, and process optimization

- Complete component development and testing against the defined evaluation criteria
- Assessment of product releases against acceptance criteria for the vision

The outcome of the construction phase is a product ready to put in the hands of its end users. At minimum, it consists of

- The software product integrated on the adequate platforms
- The user manuals
- A description of the current release

Milestone: Initial Operational Capability

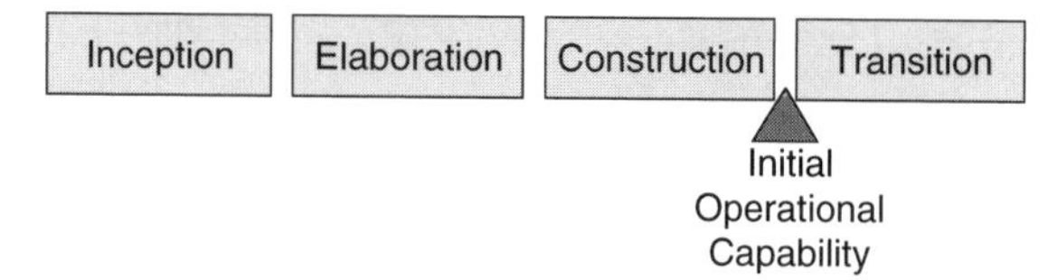

At the end of the construction phase is the third major project milestone: initial operational capability. At this point, you decide whether the software, the sites, and the users are ready to become operational without exposing the project to high risks. This release is often called a *beta* release.

The evaluation criteria for the construction phase involve answering the following questions.

- Is this product release stable and mature enough to be deployed in the user community?
- Are all stakeholders ready for the transition into the user community?
- Are the actual resource expenditures versus planned expenditures still acceptable?

Transition may have to be postponed by one release if the project fails to reach this milestone.

The Transition Phase

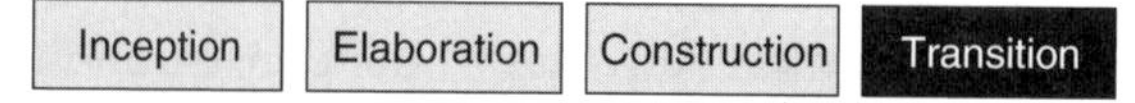

The purpose of the transition phase is to transition the software product to the user community. After the product has been given to the end user, issues usually arise that require you to develop new releases, correct some problems, or the finish features that were postponed.

The transition phase is entered when a baseline is mature enough to be deployed in the end-user domain. Typically, this means that some usable subset of the system has been completed to an acceptable level of quality and that user documentation is available so that the transition to the user will provide positive results for all parties. This phase includes the following:

- Beta testing to validate the new system against user expectations
- Parallel operation with the legacy system that the project is replacing
- Conversion of operational databases
- Training of users and maintainers
- Rollout of the product to the marketing, distribution, and sales teams

The transition phase concludes when the deployment baseline has achieved the completed vision. For some projects, this life-cycle endpoint may coincide with the starting point of the next cycle, leading to the next generation or version of the product. For other projects, it may coincide with a delivery of the artifacts to a third party responsible for operation, maintenance, and enhancements of the delivered system.

The transition phase focuses on the activities required to place the software into the hands of the users. Typically, this phase comprises several iterations, including beta releases, general availability releases, and bug-fix and enhancement releases. Considerable effort is expended in developing user-oriented documentation, training users, supporting users in the initial use of the product, and reacting to user feedback. At this point in the life cycle, however, user feedback should be confined primarily to product tuning, configuring, installation, and usability issues.

The primary objectives of the transition phase include the following:

- Achieving user self-supportability

- Achieving stakeholder concurrence that deployment baselines are complete and consistent with the evaluation criteria of the vision
- Achieving final product baseline as rapidly and cost-effectively as practical

The essential activities of the transition phase are as follows:

- Deployment-specific engineering—that is, cutover, commercial packaging and production, sales rollout, and field personnel training
- Tuning activities, including bug fixing and enhancement for performance and usability
- Assessing the deployment baselines against the vision and the acceptance criteria for the product

In the transition phase, the activities performed during an iteration depend on the goal; for fixing bugs, implementation and testing are usually enough. If new features must be added, the iteration is similar to those of the construction phase.

Depending on the type of product, this phase can range from being simple to being extremely complex. For example, a new release of an existing desktop product may be simple, whereas replacing a nation's air-traffic control system would be complex.

Milestone: Product Release

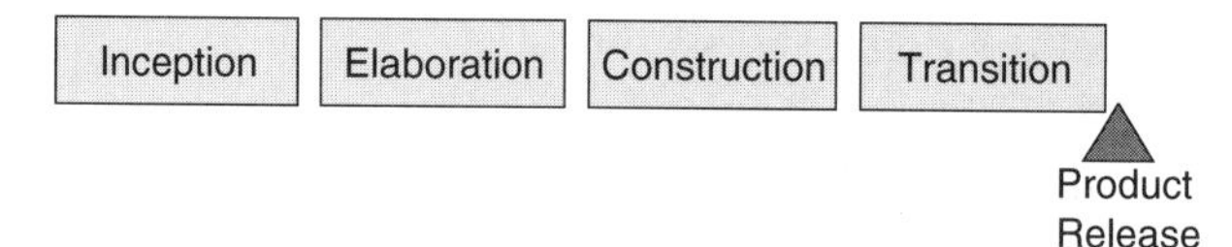

At the end of the transition phase is the fourth important project milestone: the product release milestone. At this point, you decide whether the objectives were met and whether you should start another development cycle. In some cases, this milestone may coincide with the end of the inception phase for the next cycle.

The primary evaluation criteria for the transition phase involve the answers to the following questions.

- Is the user satisfied?
- Are the actual resources expenditures versus planned expenditures still acceptable?

BENEFITS OF AN ITERATIVE APPROACH

Compared with the traditional waterfall process, the iterative process has the following advantages.

- Risks are mitigated earlier.
- Change is more manageable.
- There is a higher level of reuse.
- The project team can learn along the way.
- The product has better overall quality.

Risk Mitigation

An iterative process lets you mitigate risks earlier because integration is generally the only time that risks are discovered or addressed. As you roll out the early iterations you go through all process components, exercising many aspects of the project, including tools, off-the-shelf software, and people skills. Perceived risks will prove not to be risks, and new, unsuspected risks will be discovered.

If a project must fail for some reason, let it fail as soon as possible, before a lot of time, effort, and money are expended. Do not bury your head in the sand too long; instead, confront the risks. Among other risks, such as building the wrong product, there are two categories of risks that an iterative development process helps to mitigate early:

- Integration risks
- Architectural risks

An iterative process results in a more robust architecture because you correct errors over several iterations. Flaws are detected in early iterations as the product moves beyond inception. Performance bottlenecks are discovered at a time when they can still be addressed instead of being discovered on the eve of delivery.

Integration is not one "big bang" at the end of the life cycle; instead, elements are integrated progressively. Actually, the iterative approach that we recommend involves almost continuous integration. What used to be a lengthy time of uncertainty and pain—taking as much as 40% of the total effort at the end of a project—is now broken into six to nine smaller integrations that begin with far fewer elements to integrate.

Accommodating Changes

You can envisage several categories of changes.

Changes in Requirements

An iterative process lets you take into account changing requirements. The truth is that requirements will normally change. Requirements change and "requirements creep" have always been primary sources of project trouble, leading to late delivery, missed schedules, unsatisfied customers, and frustrated developers. But by exposing users (or representatives of the users) to an early version of the product, you can ensure a better fit of the product to the task.

Tactical Changes

An iterative process provides management with a way to make tactical changes to the product—for example, to compete with existing products. You can decide to release a product early with reduced functionality to counter a move by a competitor, or you can adopt another vendor for a given technology. You can also reorganize the contents of iteration to alleviate an integration problem that needs to be fixed by a supplier.

Technological Changes

To a lesser extent, an iterative approach lets you accommodate technological changes. You can use it during the elaboration phase, but you should avoid this kind of change during construction and transition because it is inherently risky.

Learning as You Go

An advantage of the iterative process is that developers can learn along the way, and the various competencies and specialties are more fully employed during the entire life cycle. For example, testers start testing early, technical writers write early, and so on, whereas in a non-iterative development, the same people would be waiting to begin their work, making plan after plan. Training needs, or the need for additional (perhaps external) help, are spotted early during assessment reviews.

The process itself can also be improved and refined along the way. The assessment at the end of an iteration looks at the status of the project from a product/schedule perspective and analyzes what should be changed in the organization and in the process so that it can perform better in the next iteration.

Increased Opportunity for Reuse

An iterative process facilitates reuse of project elements because it is easier to identify common parts as they are partially designed or implemented instead of identifying all commonality in the beginning. Identifying and developing reusable parts is difficult. Design reviews in early iterations allow architects to identify unsuspected potential reuse and to develop and mature common code in subsequent iterations. It is during the iterations in the elaboration phase that common solutions for common problems are found and patterns and architectural mechanisms that apply across the system are identified. For more about this issue, see Chapter 5.

Better Overall Quality

The product that results from an iterative process will be of better overall quality than in a conventional sequential process. The system has been tested several times, improving the quality of testing. The requirements have been refined and are more closely related to the users' real needs. At the time of delivery, the system has been running longer.

SUMMARY

- The sequential (or waterfall) process is fine for small projects that have few risks and use a well-known technology and domain, but it cannot be stretched to fit projects that are long or involve a high degree of novelty or risk.
- An iterative process breaks a development cycle into a succession of iterations. Each iteration looks like a miniwaterfall and involves the activities of requirements, design, implementation, and assessment.
- To control the project and to give the appropriate focus to each iteration, a development cycle is divided into a sequence of four phases that partition the sequence of

iterations. The phases are inception, elaboration, construction, and transition.

- The iterative approach accommodates changes in requirements and in implementation strategy. It confronts and mitigates risks as early as possible. It allows the development organization to grow, to learn, and to improve. It focuses on real, tangible objectives.

Chapter 5

An Architecture-centric Process

THIS CHAPTER DEFINES architecture and explains why it plays a central role in the Rational Unified Process.

THE IMPORTANCE OF MODELS

A large part of the Rational Unified Process focuses on modeling. Models help us understand and shape both the problem and the solution. A model is a simplification of reality that helps us master a large, complex system that cannot be easily comprehended in its entirety. The choice of models and the choice of techniques used to express them have a profound effect on the way we think about the problem and shape the solution. The model is not the reality ("the map is not the territory"[1]), but the best models are the ones that stick very close to reality.[2]

No single model is sufficient to cover all aspects of software development. We need multiple models to address different concerns. These models must be carefully coordinated to ensure that they are consistent and not too redundant.

1. "The map is not the territory" is fundamental to the book *Language in Thought and Action* by S.I. Hayakawa, first published in 1939. New York: Harcourt-Brace.

2. Grady Booch, James Rumbaugh, and Iver Jacobson, *The Unified Modeling Language Users Guide*. Reading, MA: Addison Wesley Longman, 1999.

ARCHITECTURE

Models are complete, consistent representations of the system to be built. The models of complex systems can be very large.

Suppose you are given the task of describing a system so that designers, programmers, users, and managers would be able to do the following:

- Understand what the system does
- Understand how the system works
- Be able to work on one piece of the system
- Further extend the system
- Reuse part of the system to build another one

Now assume that you are given only a limited amount of space for this task (for example, a maximum of 60 pages). What you would end up with is a description of the *architecture* of the system. As someone once told me, "Architecture is what remains when you cannot take away any more things and still understand the system and explain how it works."

THE IMPORTANCE OF ARCHITECTURE

For many years, software designers have had the strong feeling that software architecture was an important concept. But this concept was not very well exploited, for a number of reasons.

- The purpose of an architecture was not always well articulated.
- The concept remained fuzzy, trapped somewhere between top-level design, system concept, and requirements.
- There was no accepted way to represent an architecture.
- The process by which an architecture came into life was not described and always seemed to be some kind of art form or black magic.

Architecture ended up taking the form of an unmaintained document containing a few diagrams made of boxes and arrows reflecting imprecise semantics that could be interpreted only by its authors. But many software systems weren't very complex, so the architecture could remain an implicit understanding among software developers.

However, as systems evolve and grow to accommodate new requirements, things break in a strange fashion and the systems do not scale up. Integrating new technologies requires that we completely rebuild the systems. Systems, in other words, are not very resilient in response to changes. Moreover, the designers lack the intellectual tools to reason about the parts of the system. It is little wonder that poor architectures (together with immature processes) are often listed as reasons for project failures. Not having an architecture, or using a poor architecture, is a major technical risk for software projects.

Architecture Today

Now that all the simple systems have been built, managing the complexity of large systems has become the number one concern of software development organizations. They want their systems to evolve rapidly, and they want to achieve large-scale reuse across families of systems, building systems from ready-made components. Software has become a major asset, and organizations need conceptual tools to manage it.

The word *architecture* is now used everywhere, reflecting a growing concern and attention, but the variety of contexts in which the word is used suggests that it has not necessarily become a well-mastered concept.

For an organization to adopt an architecture focus, three things are required.

- *An understanding of the purpose*
  Why is architecture important? What benefits can we gain from it? How can we exploit it?
- *An architectural representation*
  The best way to make the concept of architecture less fuzzy is to agree on a way to represent it so that it becomes something concrete that can be communicated, reviewed, commented, and improved systematically.

- *An architectural process*
  How do you create and validate an architecture that satisfies the needs of a project? Who does it? What are the artifacts and the quality attributes of this task?

The Rational Unified Process contains some answers to all three points. But let's start by defining more precisely what we mean by software architecture, or rather, by the architecture of a software-intensive system.

A DEFINITION OF ARCHITECTURE

Many definitions of architecture have been proposed. The Rational Unified Process defines architecture as follows.[3] Architecture encompasses significant decisions about

- The organization of a software system
- The selection of structural elements and their interfaces by which the system is composed, together with their behavior as specified in the collaboration among those elements
- The composition of these elements into progressively larger subsystems
- The architectural style that guides this organization, these elements and their interfaces, their collaborations, and their composition

Software architecture is concerned not only with structure and behavior but also with context: usage, functionality, performance, resilience, reuse, comprehensibility, economic and technological constraints and trade-offs, and aesthetics.

This definition is long, but it attempts to capture the richness, the complexity, and the multiple dimensions of the concept. We can elaborate on some of the points.

Architecture is a part of design; it is about making decisions about how the system will be built. But it is not all of the design. It

3. This definition evolved from one given several years ago by Mary Shaw and David Garlan of Carnegie-Mellon University. See their textbook *Software Architecture—Perspectives on an Emerging Discipline*. Upper Saddle River, NJ: Prentice Hall, 1996.

stops at the major elements—in other words, the elements that have a pervasive and long-lasting effect on the qualities of the system, namely its evolvability and its performance.

Architecture is about structure and about organization, but it is not limited to structure. It also deals with behavior: what happens at the joints, at the seams, and across the interfaces.

Architecture not only looks inward but also looks at the "fit" of the system in two contexts: the operational context (its end users) and the development context (the organization that develops the system). And it encompasses not only a system's technical aspects but also its economic and sociological aspects.

Architecture also addresses "soft" issues such as style and aesthetics. Can an architecture be pleasing? Yes, to the educated eye it can be. Issues of aesthetics have their place in making a design uniform, easy to understand, and easy to evolve, with minimal surprises for the designers.

ARCHITECTURE REPRESENTATION

In one form or another, many different parties are interested in architecture:

- The system analyst, who uses it to organize and articulate the requirements and to understand the technological constraints and risks
- End users or customers, who use it to visualize at a high level what they are buying
- The software project manager, who uses it to organize the team and plan the development
- The designers, who use it to understand the underlying principles and locate the boundaries of their own design
- Other development organizations (if the system is open), which use it to understand how to interface with it
- Subcontractors, who use it to understand the boundaries of their chunk of development
- Architects, who use it to reason about evolution or reuse

To allow the various stakeholders to communicate, discuss, and reason about architecture, we must have an architectural representation that they understand. The architecture and its representation are not quite the same thing: by choosing a representation, we omit some of the most intangible or soft aspects.

The various stakeholders have differing concerns and are interested in different aspects of the architecture. Hence, a complete architecture is a multidimensional thing. *Architecture is not flat.*

Multiple Views

For a building, different types of blueprints are used to represent different aspects of the architecture:

- Floor plans
- Elevations
- Blueprints for electrical cabling
- Blueprints for water pipes, central heating, and ventilation
- Perspective sketches showing the look of the building in its environment

Over decades, blueprints have evolved into standard forms so that each person involved in the design and construction of the building understands how to read and use them. Each of these blueprints tries to convey one aspect of the architecture for one category of stakeholders, but the blueprints are not completely independent of each other. On the contrary, they must be carefully coordinated.

Similarly, in the architecture of a software-intensive system, you can envisage various blueprints for various purposes:

- To address the logical organization of the system
- To organize the functionality of the system
- To address the concurrency aspects
- To describe the physical distribution of the software on the underlying platform

And there are many more. These are what we call *architectural views*.[4] An architectural view is a simplified description (an abstraction) of a system from a particular perspective or vantage point, cov-

---

4. This multiple view approach is to conform to the future *IEEE Recommended Practice for Architectural Description*. See Draft 3.0 of IEEE P1471, May 1998.

ering particular concerns and omitting entities that are not relevant to this perspective.

For each view, we need to clearly identify the following:

- The point of view—the concerns and the stakeholders to be addressed
- The elements that will be captured and represented in the view, together with their relationship
- The organizational principles used to structure the view
- The way elements of this view are related to those of other views
- The best process to use in creating this view

The 4+1 View Model of Architecture

As shown in Figure 5-1, the Rational Unified Process suggests a five-view approach.[5] The following sections summarize the five views.

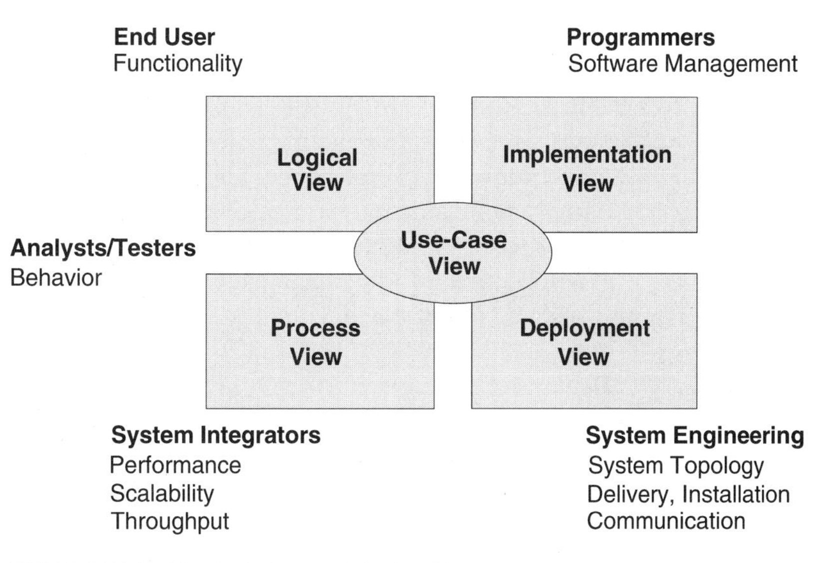

FIGURE 5-1 *The 4+1 view model of architecture*

5. Philippe Kruchten, "The 4+1 View of Architecture," *IEEE Software,* 12(6) Nov. 1995, pp. 45–50.

The Logical View

This view of the architecture addresses the functional requirements of the system—in other words, what the system should do for its end users. It is an abstraction of the design model and identifies major design packages, subsystems, and classes.

Examples include a flight, a flight plan, an airway, an airport, and an airspace package.

The Implementation View

This view describes the organization of static software modules (source code, data files, components, executables, and other accompanying artifacts) in the development environment in terms of packaging and layering and in terms of configuration management (ownership, release strategy, and so on). It addresses the issues of ease of development, management of software assets, reuse, subcontracting, and off-the-shelf components.

Examples include the source code for a flight class and the library of code for an airspace database.

The Process View

This view addresses the concurrent aspect of the system at runtime—tasks, threads, or processes as well as their interactions. It addresses issues such as concurrency and parallelism, system startup and shutdown, fault tolerance, and object distribution. It deals with issues such as deadlock, response time, throughput, and isolation of functions and faults. It is concerned with scalability.

Examples are a flight management process, flight plan entry processes, and an airspace management process.

The Deployment View

This view shows how the various executables and other runtime components are mapped to the underlying platforms or computing nodes. It is here that software engineering meets system engineering. It addresses issues such as deployment, installation, and performance.

For example, the flight management process runs on the main flight processor, whereas the flight plan entry processes run on any workstations.

The Use-Case View

This view plays a special role with regard to the architecture. It contains a few key scenarios or use cases. Initially, these are used to drive the discovery and design of the architecture in the inception and elaboration phases, but later they will be used to validate the different views. These few scenarios act as an illustration in the software architecture document of how the other views work.

Examples are the entry of a flight plan and the handover of responsibility to another air-traffic controller.

Models and Views

The preceding description of the architecture is true also for the models introduced in the first section of this chapter. We create different models to address different needs. But the models are complete representations of the system, whereas an architectural view focuses only on what is architecturally significant. If we do not make this distinction, there is no clear way to distinguish "design" from "architecture." *Not all design is architecture.*

What is "architecturally significant"? An architecturally significant element has a wide impact on the structure of the system and on its performance, robustness, evolvability, and scalability. It is an element that is important for understanding the system.

Architecturally significant elements include the following:

- Major classes, in particular those classes modeling major business entities
- Architectural mechanisms that give behavior to these classes, such as persistency mechanisms and communication mechanisms
- Patterns and frameworks
- Layers and subsystems
- Interfaces
- Major processes, or threads of control

Architectural views are like slices cut through the various models, illuminating only the important, significant elements of the models. Table 5-1 summarizes the relationship between models and views. The views useful for a system are captured in an important artifact: the software architecture description.

TABLE 5-1 *The Relationship Between Models and Views*

Model	Architectural View
Design model	Logical view
Design model*	Process view
Implementation model	Implementation view
Deployment model	Deployment view
Use-case model	Use-case view

*or a process model for a complex system

Architecture Is More Than a Blueprint

Architecture is more than just a blueprint of the system to be developed. To validate the architecture and to assess its qualities in terms of feasibility, performance, flexibility, and robustness, we must build it.

Building the architecture, validating it, and then baselining it are the number one objectives of the elaboration phase. Therefore, in addition to a software architecture description, the most important artifact associated with the architecture is an architectural prototype that implements the most important design decisions sufficiently to validate them—that is, to test and measure them. This prototype is not a quick-and-dirty throwaway prototype, but it will evolve through the construction phase to become the final system. See the section titled Prototypes in Chapter 11.

AN ARCHITECTURE-CENTRIC PROCESS

After an organization agrees on a representation of the architecture that is suitable for the problem at hand, the next issue is to master an architectural design process.

The Rational Unified Process defines two primary artifacts related to architecture:

- The software architecture description (SAD), which describes the architectural views relevant to the project
- The architectural prototype, which serves to validate the architecture and serves as the baseline for the rest of the development

These two key artifacts are the root of several others.

- Design guidelines are shaped by some of the architectural choices made and reflect the use of patterns and idioms.
- The product structure in the development environment is based on the implementation view.
- The team structure is based on the structure of the implementation view.

The Rational Unified Process defines a Worker: Architect, who is responsible for the architecture. Architects, however, are not the only ones concerned with the architecture. Most team members are involved in the definition and the implementation of the architecture, especially during the elaboration phase.

- Designers focus on architecturally significant classes and mechanisms rather than the details of the classes.
- Integrators integrate major software components, even if their implementation is very rudimentary, to verify the interfaces. Integrators focus mainly on removing integration risks related to major off-the-shelf or reused components.
- Testers test the architectural prototype for performance and robustness.

During the construction phase, the focus shifts to adding the meat and skin to the architectural skeleton. Activities reflect an ongoing concern for the architecture: tuning it, refining it, and making sure that no new design decision is introduced that would weaken or break it.

The bulk of the activities related to architectural design are described in the analysis and design workflow (see Chapter 10), but it spills over to the requirements workflow, the implementation workflow, and the project management workflow.

THE PURPOSE OF ARCHITECTURE

Now that we have embraced architecture, we can revisit its purpose. Why should an organization focus on architecture? Architecture is important for several reasons that are discussed in the following sections.

Intellectual Control

Architecture lets you gain and retain intellectual control over the project, to manage its complexity, and to maintain system integrity.

A complex system is more than the sum of its parts and more than a succession of independent tactical decisions. It must have a unifying, coherent structure so as to organize the parts systematically. It must also provide precise rules about how to grow the system without having its complexity multiply beyond human understanding.

Maintaining the system's integrity has long been the objective of system architects. Integrity, in its original meaning is "being one." We want control of the architecture to design and build systems that are almost continuously as "one" without resorting to difficult and painful "integration phases" in which all kinds of mismatched parts must be force-fitted together.

The architecture establishes a means for improved communication and understanding throughout the project by establishing a common set of references and a common vocabulary for discussing design issues. Each designer and developer will understand the context and boundary of the part he or she is working on.

Reuse

Architecture provides an effective basis for large-scale reuse.

By clearly articulating the major components and the critical interfaces between them, an architecture lets you reason about reuse. It assists both internal reuse—the identification of common parts—and external reuse—the incorporation of ready-made, off-the-shelf components.

Architecture also facilitates reuse on a larger scale: the reuse of the architecture itself in the context of a line of products that addresses varying functionality in a common domain. Architects will understand the limits of scalability, identify the generic parts, and define the variation points.

Basis for Development

Architecture provides a basis for project management.

Planning and staffing are organized along the lines of major components: layers and subsystems. Fundamental structural decisions are made by a small, cohesive architecture team; the decisions are not distributed. Development is partitioned across a set of small teams, each of which is responsible for one or several parts of the system. Different development organizations will understand which interfaces are at their disposal and will understand the limits of their variation.

Architecture and the work done during the elaboration phase provide the basis for further development, including the design guidelines, principles, styles, patterns, and mechanisms to reuse.

COMPONENT-BASED DEVELOPMENT

The Rational Unified Process supports component-based development (CBD), which is the creation and deployment of software-intensive systems that are assembled from components, as well as the development and harvesting of such components.

Component-based development is about building quality systems that satisfy business needs quickly, preferably by using parts rather than handcrafting every individual element. It involves crafting the right set of primitive components from which to build families of systems, including the harvesting of components. Some components are intentionally made; others are discovered and adapted.

A definition of *component* must be broad enough to address conventional components (such as COM/DCOM, CORBA, and JavaBeans components) as well as alternative ones (Web pages, database tables, and executables using proprietary communication). At the same time, it shouldn't be so broad as to encompass every possible artifact of a well-structured architecture.

A Definition of Component

A component is a nontrivial, nearly independent, and replaceable part of a system that fulfills a clear function in the context of a well-defined architecture. A component conforms to and provides the physical realization of a set of interfaces.

Note that component and architecture are two intertwined concepts: the architecture identifies components, their interfaces, and their interactions along several dimensions, and components exist only relative to a given architecture. You cannot mix and match your chosen components if they have not been made to fit.

There are several perspectives on components.

- *Runtime components*
  The things that are delivered, installed, and run, such as executables, processes, and dynamic link libraries (DLLs). They live at runtime on the deployment platform.
- *Development components, as seen from the development organization point of view*
  Implementation subsystems that have high internal cohesion and low external coupling and are reusable by other developers. There may not always be a one-to-one relationship between runtime components and development components, but they provide a useful "first cut" at the runtime components.
- *Business components*
  Cohesive sets of runtime components (or development components) that fulfill a large chunk of business-level functionality and are units of sale, release, or upgrade.

OTHER ARCHITECTURAL CONCEPTS

The following sections describe several other topics related to software architecture.

Architectural Style

A software architecture, or an architectural view, may have an attribute called *architectural style*, which reduces the set of possible forms to choose from and imposes a certain degree of uniformity on the architecture. The style may be defined by the selection of an architectural framework, by middleware, by a recommended set of patterns, or by an architecture description technique or tool.

Examples of architectural styles are pipe-and-filter, client-server, and event-driven styles.

Architectural Mechanism

An *architectural mechanism* is a class, a group of classes, or a pattern that provides a common solution to a common problem. It is used to give life—that is, behavior—to other classes. Mechanisms are found primarily in the middle and lower layers of an architecture. They are analogous to the standard heating conduits and plumbing fixtures that are available to the architect of a building. They provide a means of rapidly implementing desired solutions.

Examples of architectural mechanisms include a database management system (DBMS), an event broadcasting system, and a transaction server.

Architectural Pattern

A *pattern* addresses and presents a solution to a recurring design problem that arises in specific design situations. Patterns document existing, well-proven design experience. They identify abstractions that are above the level of classes, instances, and components. They provide a common vocabulary and understanding of design principles and are therefore a means of documenting software architectures. Patterns are larger-scale than mechanisms are; patterns describe broad interactions of abstract design elements that help the architect and designers think about a complex problem in an intuitive shorthand.

Software architectural patterns are analogous to patterns of architecture in buildings. When someone says that a building is a skyscraper, immediately a mental picture of the kind of building and the kind of project leaps to mind. We know that the structure will be made of steel and concrete and not of wood. The design will include elevators in addition to stairways.

Patterns and mechanisms provide the architect with a growing toolkit for solving architectural problems. Just as building architects have evolved a representation and symbology of the various problems they encounter, so too have software architects evolved intellectual tools (and software tools) to manage the complexity of systems. Patterns and mechanisms play a growing part in this toolkit.[6]

Examples of architectural patterns are Model-View-Controller (MVC) and Object Request Broker (ORB).

---

6. See Frank Buschmann, Régine Meunier, Hans Rohnert, Peter Sommerlad, and Michael Stahl, *Pattern-Oriented Software Architecture: A System of Patterns*. New York: John Wiley and Sons, 1996.

SUMMARY

- System architecture is used in the Rational Unified Process as a primary artifact for conceptualizing, constructing, managing, and evolving the system under development.
- Architecture is a complex concept that is best represented by multiple, coordinated architectural views.
- An architectural view is an abstraction of a model that focuses on its structure and its essential elements.

See the Bibliography for further readings on software architecture.

Chapter 6

A Use-Case-Driven Process

THIS CHAPTER INTRODUCES the concepts of use case, actor, and scenario. It shows how use cases are used throughout the development cycle as drivers for many activities, flowing information through several models and encouraging consistency across these models.

MODELS REVISITED

A large part of the Rational Unified Process focuses on modeling. As we explained in Chapter 5, models help us to understand and shape both the problem and the solution. The choice of models and the choice of techniques used to express them have a significant impact on the way we think about the problem and try to shape the solution.

In this chapter we focus primarily on one way to understand and model the problem. There are many ways to model the problem and to express the requirements and constraints on the system to be developed. But we must keep in mind that after this stage, we will have to formulate a model of the solution. If the model of the problem is far from the model of the solution, a great deal of effort will be expended to translate the problem from a form understandable by end users to a form understandable by designers and builders. This means that there are many more opportunities for misinterpretation, and it makes it difficult to validate the solution with respect to the stated problem. Moreover, the process involves several models that must be kept consistent.

To model the problem, the Rational Unified Process recommends the technique of use-case modeling. Use cases provide a means of expressing the problem in a way that is understandable by a wide range of stakeholders: users, developers, and acquirers.[1]

DEFINITIONS

The Rational Unified Process defines two key concepts: use case and actor.

- A *use case* is a sequence of actions a system performs that yields an observable result of value to a particular actor.
- An *actor* is someone or something outside the system that interacts with the system.

Therefore, we have the system under consideration, which is surrounded by actors (people or other systems) that interact with it, and we have use cases that define these interactions.

In reviewing these definitions you should consider several key items.

- *Actions*
  An action is a computational or algorithmic procedure that is invoked when the actor provides a signal to the system or when the system gets a time event. An action may imply signal transmissions to either the invoking actor or other actors. An action is atomic, which means it is performed either entirely or not at all.
- *A sequence of actions*
  The sequence referred to in the definition is a specific flow of events through the system. Many flows of events are pos-

1. Use cases were introduced by Ivar Jacobson in *Object-Oriented Software Engineering: A Use-Case-Driven Approach*. Reading, MA: Addison-Wesley, 1992. See also the more recent book: Ivar Jacobson, Grady Booch, and James Rumbaugh, *The Unified Software Development Process*. Reading, MA: Addison Wesley Longman, 1998.

sible, and many of them may be very similar. To make a use-case model understandable, you group similar flows of events into a single use case.

- *The system performs*
  This means that we are concerned with what the system does in order to perform the sequence of actions. The use case helps us to define a firm boundary around the system and what it does to separate it from the outside world. In this way, it helps us bound the scope of the system.
- *An observable result of value*
  The sequence of actions must yield something that has value to an actor of the system. An actor should not have to perform several use cases in order to achieve something useful. Focusing on useful value provided to an actor ensures that the use case has relevance and is at a level of granularity that can be understood by the user.
- A *particular actor*
  Focusing on a particular actor forces us to isolate the value provided to specific groups of users of the system, ensuring that the system does what they need it to. It avoids building use cases that are too big. It also prevents us from losing focus and building systems that try to satisfy the needs of everyone but in the end satisfy no one.

The description of a use case defines what happens in the system when the use case is performed. The functionality of a system is defined by a set of different use cases, each of which represents a specific flow of events.

Example of Use Cases

A bank client, for example, can use an automated teller machine (ATM) to withdraw money, transfer money, or check the balance of an account. These capabilities can be represented by a set of use cases, as shown in Figure 6-1.

Each use case represents something that the system does that provides value to the bank's customer, the Client. The collected use cases constitute all the possible ways of using the system. You can get an idea of a use-case task simply by observing its name.

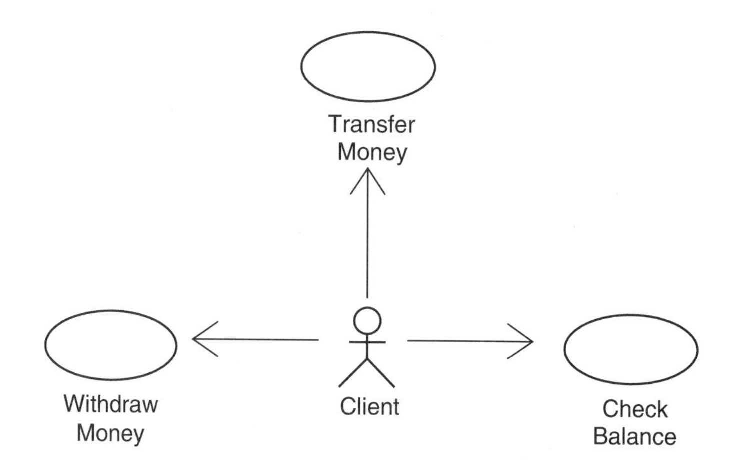

FIGURE 6-1 *Use cases for an ATM*

FLOW OF EVENTS OF A USE CASE

The most important part of the use case in the requirements workflow is its *flow of events*. The flow of events describes the sequence of actions. It is written in natural language, in a simple, consistent prose, with a precise use of terms that draws on a common glossary of the problem domain.

Let's look at an example. An initial outline of the flow of events of the use case Withdraw Money could be written as follows.

1. The use case begins when the Client inserts an ATM card. System reads and validates information on the card.

2. System prompts for PIN. Client enters PIN. The system validates the PIN.

3. System asks which operation the client wishes to perform. Client selects "Cash withdrawal."

4. System requests amounts. Client enters amount.

5. System requests account type. Client selects the account type (checking, saving, credit).

6. The system communicates with the ATM network to validate account ID, PIN, and availability of the amount requested.

7. The system asks the client whether he or she wants a receipt. This step is performed only if there is paper left to print the receipt.

8. System asks the client to withdraw the card. Client withdraws the card. (This is a security measure to ensure that Clients do not leave their cards in the machine.)

9. System dispenses the requested amount of cash.

10. System prints receipt.

11. The use case ends.

If you try to describe what the system performs by giving complete sequences of actions, you rapidly realize that there are different courses of actions—various paths through the use-case flow of events. There are alternative branches and different cases (scenarios) to consider, and different values or effects are produced. Collectively, the use-case flow of events eventually describes all these possible alternative courses.

The path chosen depends on such things as the following.

- *Input from an actor*
  The actor can decide, from several options, what to do next. For example, the bank's clients may cancel the transaction at any time, or they may decide they do not need a receipt.
- *The internal state of the system*
  For example, the ATM may be out of cash or have no paper for receipts, or the cash dispenser or printer may jam.
- *Time-outs and errors*
  For example, if the ATM user does not respond within a specified time interval, the system might automatically cancel the transaction. If the user enters the wrong PIN several times, the transaction might be canceled and the card confiscated.

We do not want to have to express each possible alternative flow in a separate use case; rather, we want to group them with any related use-case flow of events. Such grouping defines a use-case *class*. Each instance of a use-case class is a specific flow of events, or a specific path, through the use case. Instances of use-case classes are also called *scenarios*. But it is usually sufficient and not ambiguous to call a use-case class a use case, and a use-case instance a scenario.

Scenarios are used in the process to extract and emphasize one particular sequence of actions. Also, when you're trying to find use cases during the early stages of a project, it is easier to start from a specific scenario and then expand it to more flows of events, finally generalizing it to a full-blown use case.

USE-CASE MODEL

The use-case model is a model of the system's intended functions and its environment, and it serves as a contract between the customer and the developers. It comprises a set of all use cases for the system, together with the set of all actors, so that all functionality of the system is covered.

The use-case model for the ATM example we discussed earlier could be completed with use cases for the bank staff and use cases for the installation personnel. The use-case model is complemented by nonfunctional specifications covering some of the aspects that cannot be easily conveyed by a use case or that apply to many or all use cases. Examples are high-level product requirements, development constraints, safety or security attributes, and so on (see Chapter 9).

As the defined system grows larger, it is useful to make sure that certain terms are used consistently throughout all use cases. To address this need, we create as a companion document a project *glossary* or, even better, a simple object model of the domain called a *domain model* (see Chapter 8).

USE CASES IN THE DEVELOPMENT PROCESS

The Rational Unified Process is a use-case-driven approach. This means that the use cases defined for a system are the basis for the entire development process (see Figure 6-2).

The use-case model is a result of the requirements workflow. In this front-end activity, the use cases are used to capture what the system should do from the user's point of view. Thus, use cases act as the common language for communication between the customer or users and the system developers.

In analysis and design, use cases are the bridge that unites requirements and design activities. They serve as the basis for use-case realizations, which describe how the use case is performed in terms of interacting objects in the design model. The objects and

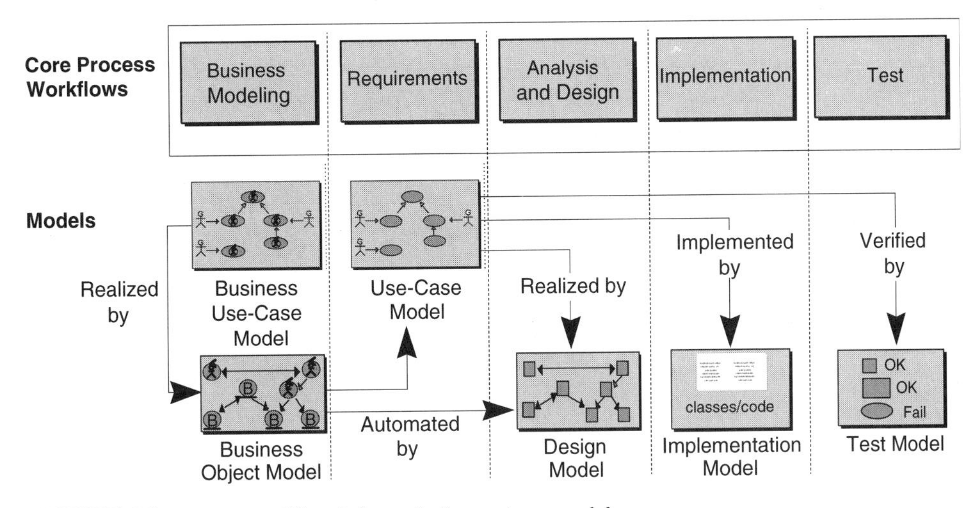

FIGURE 6-2 *Use cases "flow" through the various models*

classes are most likely found by walking through the use cases. This technique ensures that all the required behavior is represented in the system design.

During implementation, the design model is the implementation specification. Because use cases are the basis for the design model, they are implemented in terms of design classes. Use-case realizations in the design model are used to understand the dynamics of the system and determine where to optimize for performance.

During testing, use cases constitute the basis for identifying test cases and test procedures. In other words, each use case is performed to verify the system.

Other activities are also attached to the use cases. For example, because the use cases specify how an actor (a user) interacts with the system, use cases provide much of the structure and contents for user manuals.[2]

In project management, as you will see in Chapter 7, use cases and scenarios are used to define the contents of iterations. Estimates of the effort can be derived from the use-case description by techniques such as function point analysis. In deployment, use-case packages can serve to plan a phased deployment or to define

2. Conversely, when you're trying to rapidly put a use-case model in place for a legacy system, consider using the user manuals as a starting point.

system variants. Definition and prototyping of user interfaces can be derived from use cases in the form of use-case storyboards.

The use-case flow of events expresses the behavior of the system in a "gray box view" of the system, whereas *use-case realization* in design is the "white box view." It shows how the use case is actually performed in terms of interacting objects and classes.

Optionally, in business modeling, we use the same concept of use case but at the level of the whole business rather than only the system under consideration. The business use-case model describes high-level business processes and provides the context and the source of information for expressing the system's use cases. We describe this in Chapter 8, The Business Modeling Workflow.

THE SCOPE OF A USE CASE

It is often difficult to decide whether a set of user-system interactions, or *dialog*, is one or several use cases. Consider the use of an ATM: you can walk up to the cash dispenser, insert your bank card, punch in your PIN, and then withdraw money, deposit money, check your balances, or transfer funds between accounts. Then you can get a receipt for your transaction and get your card back. (For the moment, let's forget about multiple transactions.)

Is inserting your ATM card, entering your PIN, and having it validated a use case? Does it provide value to you? Suppose you walked up to an ATM, inserted your card, and entered your PIN, and the machine told you that you correctly entered your PIN and then gave you your card back? Would you be happy? Of course not! We use the ATM to get money, transfer funds, and so on. Those are the things of value that the ATM does; those are its use cases: withdraw cash, transfer funds, deposit funds, and inquire into balances. Each use case includes the complete dialog, from inserting the ATM card to selecting the transaction type to getting the receipt and the return of the card.

Use cases emerge when you focus on the things of value that a system provides to an actor and when you group the sequences of actions that a system takes to provide those things of value.

You want to keep the actions together so that you can review them at the same time, modify them together, test them together, write manuals for them, deploy them together, and, in general, manage them as a unit. The need for this approach becomes obvious in larger systems. It is different from a "functional decomposition" approach, in which you rapidly lose track of the value and purpose of each little bit of functionality. The reason for grouping pieces of functionality and calling it a use case is that you want to manage these pieces of functionality together throughout the life cycle.

HOW A USE CASE EVOLVES

In early iterations, during the elaboration phase, only a few use cases (those that are considered architecturally significant) are described in detail beyond the brief description. You should always start by developing an outline of the use case (in a step-by-step format) before delving into the details. Toward the end of elaboration, all use cases that you plan to describe in detail should be completed.

Your model will often contain use cases that are so simple that they do not need a detailed description of the flow of events, and a step-by-step outline is sufficient. The criterion for making this decision is that there shouldn't be any disagreement among the various stakeholders about what the use case means and that designers and testers are comfortable with the level of detail provided by the step-by-step format. Examples are use cases that describe simple entry or retrieval of data from the system.

ORGANIZING USE CASES

A small system can be expressed as half a dozen use cases involving two or three actors. As you tackle larger systems, you must define structuring and organizing principles. Otherwise, you sink under a profusion of use cases, some of which have in common major parts of the flow of events. This profusion would make the requirements hard to understand and would make activities such as planning, prioritizing, and estimating much more difficult.

The first structuring concept is that of a *use-case package*: you group related use cases in the same container. You can also exploit relationships between use cases. To achieve this, you must look more closely at the flows of events.

You can view a flow of events as several subflows—one main flow and one or more alternative flows—that, taken together, yield the total flow of events. You can then reuse the description of a subflow in the flows of events of other use cases: subflows in the description of one use case's flow of events may be the same as those of other use cases. In the design, you should have the same objects perform this same behavior in all the relevant use cases; that is, only one set of objects should perform this behavior no matter which use case is executing.

If the behavior is common to more than two use cases, or if it forms an independent part, the model might be clearer if you model the behavior as a use case of its own that can be reused by the original use cases. There are three ways to exploit commonality between use cases:

- Inclusion
- Extension
- Generalization/specialization[3]

Several use cases can *include* a common flow-of-event text organized as a use case. This technique is similar to the factoring you do in programming by means of subprograms. In the ATM example, both Withdraw Money and Check Balance include "User enters PIN." There is no need to express the flow of events associated with the entry of the PIN number two or three times in the same model. Both use cases could include a use case "User authentication."[4]

A use case can *extend* an existing one. For example, in a telephone switch, a basic person-to-person telephone call can be extended by call forwarding or conference call functionality. The extension occurs at specific points, called *extension points*, in the extended use case. The extension can be thought of as "injecting" additional descriptive text into the extended use case, at the extension point, under particular conditions. Extension is used primarily to simplify complex flows of events, to represent optional behavior,

3. Starting with UML 1.3, the "use" relationship between use cases has been broken down between "include" and generalization/specialization; the "use" relationship has therefore been eliminated to prevent confusion.

4. This very small example may be misleading. Do not fall into the trap of doing functional decomposition, in which you break down all use cases into a myriad of little meaningless use cases that fail to be of value to any actor.

or to handle exceptions. The result of modeling with extensions is that it should make the basic flow of events of a use-case more understandable and should make the maintenance of the use case model easier over time. Unlike the earlier "include" example, in which the basic flow of events is incomplete without the inclusion, when you use "extend," the basic flow of events should still be complete, and it should not have to refer to the extending flow of events to be understandable.

A use case can *specialize* a more general one by adding or refining some of the original flow of events. For example, in an air-traffic control system, the Entry of an ICAO Flight Plan follows the same basic principles and sequences of actions as the general Entry of a Flight Plan but has some refinements as well as specific steps, or attributes. Specialization in the use-case model provides a way to model the behavior of common application frameworks and makes it easier to specify and develop variants of the system.

A common mistake is to overuse these three relationships, a practice that leads to the creation of a use-case model that is difficult to understand and maintain. It also generates a lot of overhead.

CONCURRENCY

Sometimes you may worry about interactions between use cases at runtime, such as several use cases using the same resources and thereby creating an opportunity for conflict. The purpose of the use-case model is to make sure that all functional requirements are handled by the system; to remain focused on this, we ignore (for the moment) nonfunctional requirements, including concurrency. You should assume at this stage that all use case (instances) can run concurrently without any problem. When you design the system, you should then ensure that all nonfunctional requirements are handled correctly, including ensuring that concurrency requirements are dealt with properly.

USE CASES IN THE UML

Use cases and actors are defined in the Unified Modeling Language (UML), and the Rational Unified Process uses use-case diagrams to visualize the use-case model, including all relationships between use cases.

SUMMARY

- Use cases are a means of expressing requirements on the functionality of the system.
- Written using concise, simple prose and illustrated with activity diagrams, use cases are understandable by a wide range of stakeholders.
- Use cases drive numerous activities in the Rational Unified Process:
  - The creation and validation of the design model
  - The definition of the test cases and test procedures in the test model
  - The planning of iterations
  - The creation of user manuals
  - The deployment of the system
- Use cases help synchronize the content of various models.
- A use case is managed as a unit throughout the development.
- Use cases are also used to model the business, providing a context for the system development.
- Use cases are organized in a use-case model, which also expresses the relationships between them.
- Scenarios are described instances of use cases.

Part II
Process Workflows

Chapter 7
The Project Management Workflow

THIS CHAPTER INTRODUCES the concepts of risk and metrics—two key elements in planning and controlling an iterative process. We describe how to plan an iterative process and how to decide on the duration and contents of an iteration.

PURPOSE

Software project management is the art of balancing competing objectives, managing risk, and overcoming constraints to successfully deliver a product that meets the needs of the customers (the ones who pay the bills) and the end users. The fact that few projects are 100% successful is an indicator of the difficulty of the task. Our goal with the software project management workflow of the Rational Unified Process is to make the task easier by providing guidance in this area. It is not a recipe for success, but it presents an approach to managing the project that will markedly improve the odds of delivering software successfully.

The project management workflow has the following three purposes:

- To provide a framework for managing software-intensive projects

- To provide practical guidelines for planning, staffing, executing, and monitoring projects
- To provide a framework for managing risk

However, this workflow of the Rational Unified Process does not attempt to cover all aspects of project management.[1] For example, it does not cover issues such as

- Managing people: hiring, training, coaching
- Managing budgets: defining, allocating
- Managing contracts with suppliers and customers

This workflow focuses mainly on the specific aspects of an iterative development process:

- Risk management
- Planning an iterative project through the life cycle and planning a particular iteration
- Monitoring the progress of an iterative project and metrics

PLANNING AN ITERATIVE PROJECT

It is usually not very difficult to convince a project manager of all the benefits of an iterative project (see Chapter 4). But when it comes time to define one, project managers are often perplexed: few of the traditional planning techniques seem to apply. Some new questions arise.

- How many iterations do I need?
- How long should they be?
- How do I determine the contents and objectives for an iteration?
- How do I track the progress of an iteration?

Among the objectives of project planning are the following:

- To allocate tasks and responsibilities to a team of people over time

---

1. Project managers are invited to read a companion book: Walker Royce, *Software Project Management: A Unified Framework.* Reading, MA: Addison Wesley Longman, 1998.

- To monitor progress relative to the plan and to detect potential problems as the project is rolled out

Planning also deals with managing inanimate resources, such as facilities, equipment, and budgets, but we will not cover this aspect because it is not significantly affected by the iterative approach.

Two Levels of Plans

There have been countless ambitious but doomed attempts to plan large software projects from start to finish in minute detail, as you would plan the construction of a skyscraper. As planners try to pin down activities and tasks at the person and day level several months or years in advance, I have seen the resulting Gantt charts and logic networks of activities cover the walls of several rooms and lobbies. For such plans to be realistic, you must have a very good understanding of what will be built, you must have stable requirements and a stable architecture, and you must have built a similar system from which you can derive a detailed work breakdown structure (WBS).

But how can you plan to have Joe code module GGART in week 37 if you do not even know about the existence of module GGART?

The approach works well in industries in which the WBS is more or less standard and stable and the ordering of tasks is more deterministic because of, for example, the underlying laws of physics. When constructing a building, you cannot bring up floors 1 and 4 in parallel at the same time unless you've done work on floors 2 and 3.

In an iterative process, we recommend that the development be based on two kinds of plans:

- A coarse-grained plan: the phase plan
- A series of fine-grained plans: the iteration plans

The Phase Plan

The *phase plan* is a coarse-grained plan, and there is only one per development project. It captures the overall "envelope" of the project for one cycle (and maybe the following cycles, if appropriate). It can be summarized as follows.

- *Dates of the major milestones:*
  - Life-cycle objective (end of inception, project well scoped and funded)
  - Life-cycle architecture (end of elaboration, architecture complete)
  - Initial operational capability (end of construction, first beta)
  - Product release (end of transition and of the cycle)
- *Staffing profile:* which resources are required over time
- *Dates of the minor milestones:* end of each iteration and its primary objective, if it is known

The phase plan is produced very early in the inception phase and is updated as often as necessary. It does not require more than one or two pages. It refers to the vision document to define the scope and assumptions of the project (see Chapter 9).

The Iteration Plan

An iteration plan is a fine-grained plan, and there is one per iteration. A project usually has two iteration plans "active" at one time:

- The current iteration plan (the one for the current iteration), which is used to track progress
- The next iteration plan (the one for the pending iteration), which is built toward the second half of the current iteration and is ready at the end of the current iteration

The iteration plan is built using traditional planning techniques and tools (Gantt charts and so on) to define the tasks and their allocation to individuals and teams. The plan contains important dates, such as major builds, arrival of components from other organizations, and major reviews.

You can picture the iteration plan as a window moving through the phase plan, acting as a magnifier, as shown in Figure 7-1.

Because the iterative process is dynamic and is meant to accommodate changes in goals and tactics, no purpose is served by spending an inordinate amount of time producing detailed plans that extend beyond the current planning horizon. Such plans are difficult to maintain, rapidly become obsolete, and are typically ignored by the performing organization. The iteration plan covers about the right span of time and is the right granularity to do a good job of detailed planning.

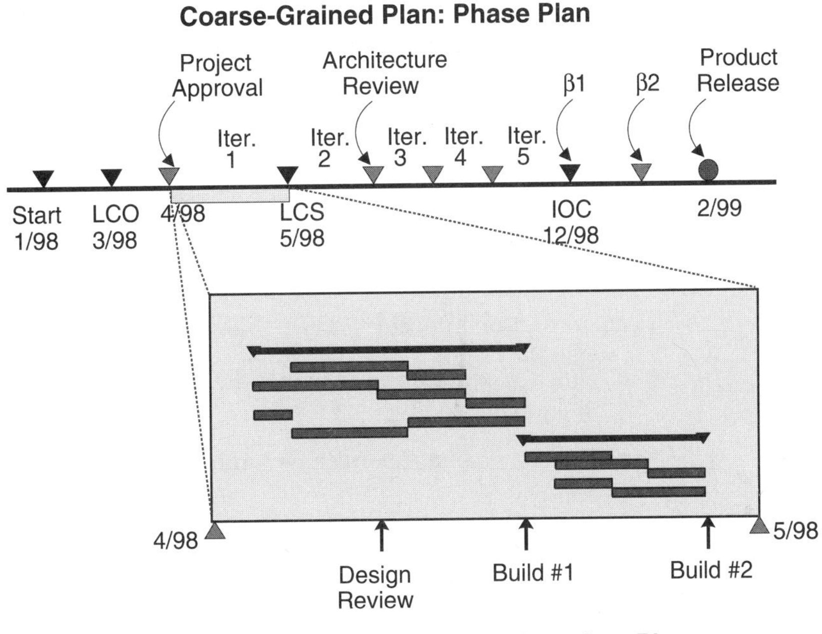

FIGURE 7-1 *Phase plan and iteration plan*

THE CONCEPT OF RISK

As Tim Lister says, "All the risk-free projects have been done."[2] The software development process primarily takes care of the *known* aspects of software development. You can precisely describe, schedule, assign, or review only what you know must be done. Risk management takes care of the *unknown* aspects. Many organizations work in a "risk denial" mode: estimating and planning proceed as if all variables were known and assume that work is mechanical and personnel are interchangeable.

What Is a Risk?

Many decisions in an iterative life cycle are driven by risks. To make effective decisions, you need a good grasp of the risks the project faces and clear strategies for mitigating or dealing with them.

2. *Software Risk Management Is Software Project Management*, Seminar at Software Productivity Center, Vancouver, BC, May 1996.

In everyday life, a risk is an exposure to loss or injury or a factor, thing, element, or course involving uncertain danger. We can define risk more specifically in software development as a variable that, within its normal distribution, can take a value that endangers or eliminates success for a project.

In plain terms, a risk is whatever may stand in our way to success and is currently unknown or uncertain. We can define *success* as meeting the entire set of all requirements and constraints held as project expectations by those in power. We can further qualify risks as direct or indirect.

- *Direct risk:* a risk over which the project has a large degree of control
- *Indirect risk:* a risk over which the project has little or no control

We can also add two important attributes:

- The probability of occurrence
- The impact on the project (severity)

These two attributes can often be combined in a single *risk magnitude indicator*, and five discrete values are sufficient: high, significant, moderate, minor, and low.

Strategies: How to Cope with Risks

The key idea in risk management is that you not wait passively until a risk materializes and becomes a problem (or kills the project) before you decide what to do with it. For each perceived risk, you must decide in advance what you are going to do. Three main routes are possible.[3]

- *Risk avoidance:* Reorganize the project so that it cannot be affected by the risk.
- *Risk transfer:* Reorganize the project so that someone or something else bears the risk (the customer, vendor, bank, or another element).

3. Barry W. Boehm, "Software Risk Management: Principles and Practice," *IEEE Software*, Jan. 1991, pp. 32–41.

- *Risk acceptance:* Decide to live with the risk as a contingency. Monitor the risk symptoms and determine what to do if the risk materializes.

When accepting a risk, you should do two things.

- *Mitigate the risk:* Take immediate, proactive steps to reduce the probability or the impact of the risk.
- *Define a contingency plan:* Determine the course of action to take if the risk becomes an actual problem; in other words, create a "plan B."

Risks play a major role in planning iterations, as you will see later.

THE CONCEPT OF METRICS

Why do we measure? We measure primarily to gain control of a project and therefore to manage it. We measure to evaluate how close or far we are from the plan's objectives in terms of completion, quality, and compliance with requirements. We also measure so that we can better plan a new project's effort, cost, and quality based on past experience. Finally, we measure to evaluate the effects of changes and assess how we have improved over time on key aspects of the process's performance (see Chapter 17).

Measuring key aspects of a project adds a non-negligible cost, so we do not measure something simply because we can. We must set precise goals for a measurement effort and collect only metrics that allow us to satisfy these goals. There are two kinds of goals.[4]

- *Knowledge goals*
  These goals are expressed by the use of verbs such as *evaluate, predict,* and *monitor.* They express a desire to better understand your development process. For example, you may want to assess product quality, obtain data to predict testing effort, or monitor test coverage or requirements changes.

4. K. Pulford, A. Kuntzmann-Combelles, and S. Shirlaw, *A Quantitative Approach to Software Management—The ami Handbook.* Reading, MA: Addison-Wesley, 1995.

- *Change or achievement goals*
  These goals are expressed by the use of verbs such as *increase, reduce, improve,* and *achieve*. They express an interest in seeing how things change or improve over time, from one iteration to another, and from one project to another.

The following are examples of goals that might be set in a software development effort.

- Monitor progress relative to the plan.
- Improve customer satisfaction.
- Improve productivity.
- Improve predictability.
- Increase reuse.

These general management goals do not readily translate into metrics. We must translate them into smaller subgoals (or *action* goals), which identify the actions that project members must take to achieve the goal. We also must make sure that the people involved understand the benefits.

For example, the goal "Improve customer satisfaction" would break down into the following action goals.

- Define customer satisfaction.
- Measure customer satisfaction over several releases.
- Verify that satisfaction improves.

The goal "Improve productivity" would include these subgoals.

- Measure effort.
- Measure progress.
- Calculate productivity over several iterations or projects.
- Compare the results.

Some of the subgoals (but not all of them) would require that you collect metrics. For example, "Measure customer satisfaction" can be derived from

- A customer survey (in which the customers would give marks for different support aspects)
- The number and severity of calls to a customer-support hotline

WHAT IS A METRIC?

There are two kinds of metrics.

- A *metric* is a measurable attribute of an entity. For example, project effort is a measure (that is, a metric) of project size. To calculate this metric you would need to sum all the time-sheet bookings for the project.
- A *primitive metric* is a raw data item that is used to calculate a metric. In the preceding example, the time-sheet bookings are the primitive metrics. A primitive metric is typically a metric that exists in a database but is not interpreted in isolation.

Each metric comprises one or more collected metrics. Consequently, each primitive metric must be clearly identified, and its collection procedure must be defined.

Metrics to support change or achievement goals are often *first-derivative* over time (or iterations or project). We are interested in a trend and not in the absolute value. If our goal is "Improve quality," we must check that the residual level of known defects diminishes over time.

WORKERS AND ARTIFACTS

In the Rational Unified Process, how is all this translated concretely in terms of workers, artifacts, activities, and workflow? Figure 7-2 shows the worker and the artifacts of the management workflow. The project manager is the only worker in this workflow.

The key artifacts of the project management workflow are as follows:

- The software development plan (SDP), which contains several artifacts:
  - Risk list
  - Project plan
  - Measurement plan
- The business case
- The iteration plans (one per iteration)
- The iteration assessment
- Any other periodic status assessment

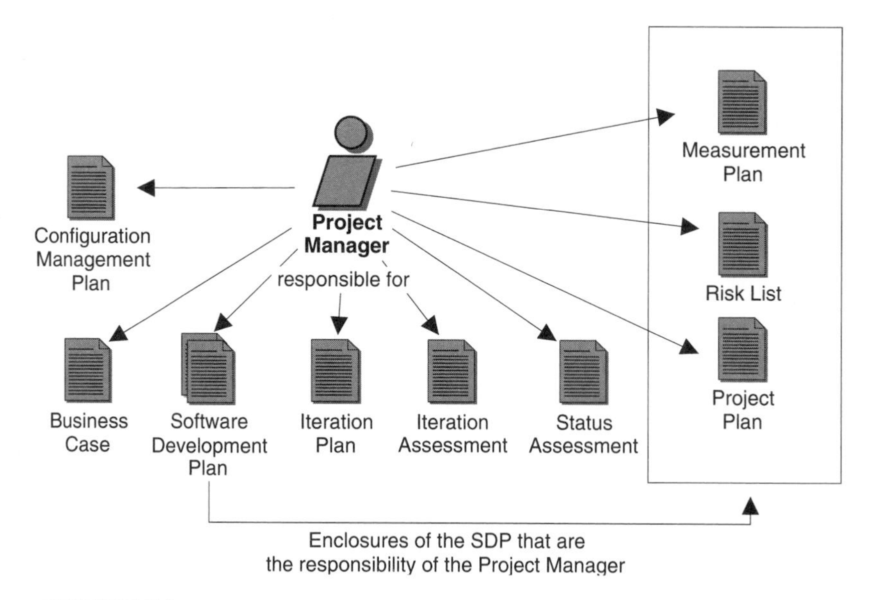

FIGURE 7-2 *Worker and artifacts in the project management workflow*

Other plans are part of the SDP, but they are developed by other workers. Here are two examples:

- Configuration management plan, developed by the Worker: Configuration Manager (see Chapter 13)
- Development case (the process used for project), developed by the Worker: Process Engineer (see Chapters 14 and 17)

WORKFLOW

Figure 7-3 shows a typical workflow in project management. In the next few sections, we will cover a few of these activities.

Building a Phase Plan

To start building a phase plan, you need at least a rough estimate of the "size of the mountain."

- How high is it (total effort)?
- When do I need to arrive at the top (date of final release)?

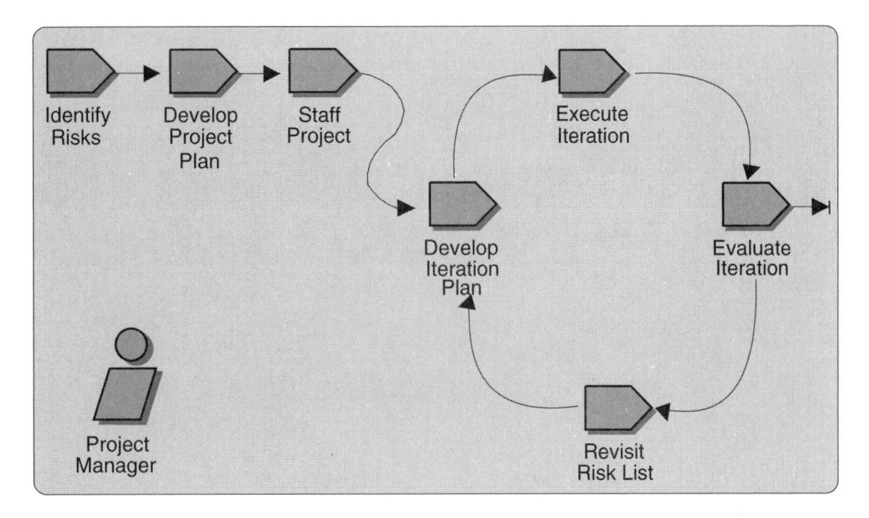

FIGURE 7-3 *A workflow in project management*

You then work backward from the end date to "plant" tentative dates for the major milestones.

Staff/Schedule/Scope Trade-off

Many studies (and common sense) have shown again and again that you cannot trade staff for schedule. This is the classic example: "If it takes nine months for a woman to make a baby, why can't we have nine women produce one in a month?" Or "Adding people to a late project delays it further." You can use a cost model, such as COCOMO, to validate that you have the right relationship between effort, elapsed time, and staffing level.

To reach a reasonable ratio in your product's first generation, you usually must trade off features, or you must be creative in other ways, such as increasing reuse. Note that you could trade off another parameter—quality—but that is another discussion.

The Rubber Profile

After you have an idea of the limits of the three variables, you must shape your effort profile and refine the dates of milestones, taking into account the specific circumstances of your project.

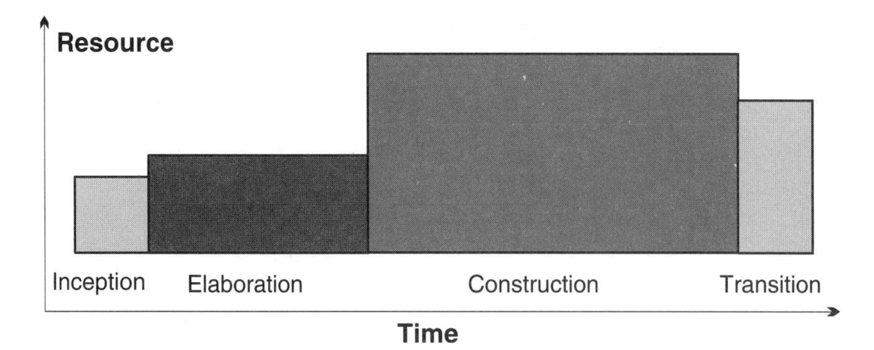

FIGURE 7-4 *Typical project profile*

To do this you can start from a typical profile. The profile in Figure 7-4 shows the relative duration and effort per phase. It is suitable for a project that has the following characteristics:

- Is of moderate size and effort
- Is in an initial development cycle
- Has no preexisting architecture
- Has a small number of risks and unknowns

Table 7-1 shows the profile in tabular fashion.

Now let's assume that this profile is made of rubber. Let's stretch it and massage it to fit your circumstances using the following heuristics.

- If you need a long time to scope the project, find the funding, conduct market research, or build an initial proof-of-concept prototype, stretch the inception phase.

TABLE 7-1 *Relative Weight of the Phases on Schedule and Effort for a Typical Project*

Phase	Schedule	Effort
Inception	10%	5%
Elaboration	30%	20%
Construction	50%	65%
Transition	10%	10%

- If you have no architecture in place, if you envisage using new and unknown technology, and if you have severe performance constraints, a number of technical risks, and a lot of new staff, lengthen the elaboration phase.
- If this is the second generation of an existing product (an evolution cycle) and if you will make no major changes to the architecture, shrink the inception and elaboration phases.
- If you must hit the market quickly—because you are late or because you are creating the market—and plan to finish the product gradually, you can shorten the construction phase and lengthen the transition phase.
- If you have a complex deployment, such as replacing an old system without interruption of service, or if you have regulatory requirements to satisfy or certification to obtain (as in domains such as medical instrumentation, nuclear industry, avionics, or public telephony), you may have to stretch the transition phase.

Again and again, verify that you are not overstaffing the project.

Another element to consider is *how many* iterations you will perform in each phase. Before deciding this, let's first discuss the issue of the duration of an iteration.

Duration of an Iteration

We have defined an iteration as a nearly complete miniproject in which you go through all major workflows, resulting, in most cases, in an executable, yet incomplete, system. Although the cycle (edit, compile, test, debug) sounds like an iteration, it is not what we have called iteration here. The daily or weekly build, in which you incrementally integrate and test increasing numbers of elements of the system, may sound like an iteration, but such builds are not what we call an iteration. An iteration starts with planning and requirements and ends with a release, either internal or external.

How quickly you can iterate depends primarily on the size of the development organization. Here are some examples.

- Five people can do some planning on a Monday morning, have lunch together every day to monitor progress, reallocate tasks, start doing a build on Thursday, and complete the iteration by Friday evening.

- With 20 people, achieving this scenario would be rather difficult. It would take more time to distribute the work, synchronize between the subgroups, and integrate. An iteration might take three or four weeks.
- With 40 people, it takes a week for the nervous influx to go from the brain to the extremities. You have intermediate levels of management, and the common understanding of the objective will require more formal documentation and more ceremony. Three months is a more reasonable iteration length.

Other factors come into play: the degree of the organization's familiarity with the iterative approach; the stability and maturity of the organization; and the level of automation used by the team to manage code (for example, distributed configuration management), distribute information (such as an internal Web), and perform testing. Also, be aware that an iteration has some fixed overhead for planning, synchronizing, and analyzing the results.

Convinced by the tremendous benefits of the iterative approach, you might be tempted to iterate furiously, but the human limits of your organization will cool your fervor.

Although this is just an empirical approach, Table 7-2 gives some order of magnitude of iteration duration we collected on a few actual iterative projects.[5]

TABLE 7-2 *Iteration Duration for a range of Iterative Projects*

Lines of Code	Number of People	Duration of an Iteration
5,000	4	2 weeks
20,000	10	1 month
100,000	40	3 months
1,000,000	150	8 months

5. Joe Marasco looked at this sample of projects and noted that the duration, D, in weeks, was related to the size of the project, S (in thousands of lines of code), by the formula $D_{weeks} = \sqrt{S_{ksloc}}$, but you should take this with a grain of salt.

Number of Iterations

After you have an idea of the kind of iteration duration that your organization can tolerate, you can complement the "rubber profile" heuristics by looking at how many iterations you should perform in each phase.

Often, in the inception phase, there will be no real iteration; no software is produced, and there are only planning and marketing activities. In some cases, however, you will have an iteration for the following:

- Building a prototype to convince yourself or your sponsor (perhaps your customer or a venture capitalist) that your idea is a good one
- Building a prototype to mitigate a major technical risk such as trying out a new technology or a new algorithm or verifying that a performance objective is attainable
- Getting your organization up to speed with tools and process

Remember that the goal of inception is not to accumulate code. After all, you want these answers quickly.

So that's zero or one iteration.

In the elaboration phase, you should plan at least one iteration. If you have no architecture to start with and must accommodate a lot of new factors—new people, tools, technology, platform, or programming language—then you should plan two or three iterations. You cannot tackle all the risks at once. You may need to show a prototype to the customer or end users to help them better define the requirements (remember the IKIWISI effect). You will need an iteration to correct your early mistakes on the architecture.

So this gives us one to three iterations.

In the construction phase, you should plan at least one iteration. Two is more reasonable if you want to exploit the benefits of iterative development and allow a better job of integration and testing. For more ambitious projects, three or more iterations are even better if the organization can support the stress and if there is a sufficient level of automation and process maturity.

So that's one to three iterations.

In the transition phase, plan at least one iteration—for example, final release after beta. Too often, the realities of the market or the (poor) quality of the initial release will force you to do more iterations.

That's one to two iterations.

So over the entire development cycle [I, E, C, T], we have three levels:

Low:	three iterations	[0, 1, 1, 1]
Typical:	six iterations	[1, 2, 2, 1]
High:	nine iterations	[1, 3, 3, 2]

We can summarize by saying that "normal" projects have 6 ± 3 iterations. We call this our "six plus or minus three" rule of thumb.

BUILDING AN ITERATION PLAN

Now let's examine how to build a fine-grained plan for one iteration, an activity that you will repeat once per iteration. We have seen that the nature and focus of iterations vary over time from phase to phase. Although the criteria that we will take into consideration will vary through the life cycle, the detailed recipe remains the same. So we take an iteration in the elaboration phase as our main example, and later we describe how iterations differ in other phases.

First, consider the iteration, its length, and the resources you have allocated to it to be the bounding box. The idea is to avoid a situation in which you first define ambitious goals and plan to achieve them, only to discover that one iteration is not enough. You must define only enough work to fill the iteration. You are scheduling by time and not by volume.

The iteration plan describes at a fine level of granularity what will happen in the iteration. It allocates activities to workers and allocates individuals to workers. (Remember that we use the term *worker* in a special sense; see Chapter 3.) A planning tool, such as Microsoft Project, is useful for handling the details, in particular the dependencies and allocation of tasks to individuals.

To build an iteration plan, follow these steps.

1. Define objective criteria for the success of the iteration. You will use these criteria at the end of the iteration, in an iteration assessment review, to decide whether the iteration was a success or a failure.
2. Identify the concrete, measurable artifacts that will need to be developed or updated and the activities that will be required to achieve this.

3. Beginning with a typical iteration work breakdown structure, massage it to take into account the actual activities that must take place.
4. Use estimates to assign duration and effort to each activity, keeping all numbers within your resource budget.

Now let's look at how an iteration varies from phase to phase.

Iteration in the Elaboration Phase

There are three main drivers for defining the objectives of an iteration in the elaboration phase:

- Risk
- Coverage
- Criticality

The main driver to determine the iteration objectives is risk. You must mitigate or retire your risks as early as you can. This is especially the case in the elaboration phase, when most of your risks should be mitigated, but it can continue to be a key driver in the construction phase as some risks remain high or new risks are discovered.

But because the goal of the elaboration phase is to baseline an architecture, other considerations must come into play, such as ensuring that the architecture addresses all aspects of the software to be developed (*coverage*). This process is important because the architecture will be used for further planning: organization of the teams, estimation of code to be developed, and product structure on the development environment.

Although focusing on risks is important, you should keep in mind the primary missions of the system. It is good to solve all the difficult issues, but it must not be done to the detriment of the core functionality: make sure that the critical functions or services of the system are covered (*criticality*) even if no perceived risk is associated with them.

Let's assume that we have a current, up-to-date list of project risks. For the most damaging risks, identify a scenario in one use case that would force the development team to "confront" the risk. The following are two examples.

1. If there is an integration risk, such as "database D working properly with OS Y," be sure to include one scenario that involves some database interaction even if it is only a modest one.
2. If there is a performance risk, such as "excessive time to compute the trajectory of the aircraft," be sure to include one scenario that includes this computation, at least for the most obvious and frequent case.

For criticality, make sure that the most fundamental function or services provided by the system are included. Select some scenarios from the use cases that represent the most common, or the most frequent, form of the service or feature offered by the system.

For example, for a telephone switch, the plain station-to-station call is the obvious "must" for an early iteration. It's far more important to get this right than to perfect the convoluted failure modes in operator configuration of the error-handling subsystem.

For coverage, toward the end of the end of the elaboration phase, include scenarios that touch areas you know will require development even if these scenarios are neither critical nor risky.

It is often economical to create long end-to-end scenarios that address multiple issues at once. For example, suppose that one single scenario is critical and confronts two technical risks. Try to achieve some balance and avoid scenarios that are too "thick" too early—that is, ones that try to cover too many different aspects, variants, and error cases.

Also, in the elaboration phase, keep in mind that some of the risks may be of a more human nature, such as managing change, fostering a team culture, training workers, and introducing new tools and techniques. Simply reviewing them during an iteration tends to mitigate these risks, but introducing too many changes at the same time becomes destabilizing.

The following are examples of iteration objectives.

1. *Create one subscriber record on a client workstation; the record is to be stored in the database on the server, including user dialog but not including all fields; assume that no errors are detected.* This scenario combines critical functions with integration risks (database and communication software) and integration issues (dealing with two different plat-

forms). It also forces designers to become familiar with a new GUI design tool. Finally, it leads to production of a prototype that can be demonstrated to end users for feedback.

2. *Make sure that as many as 20,000 subscribers can be created and that access to one subscriber takes no longer than 200 milliseconds.* This scenario addresses key performance issues (data volume and response time) that can dramatically affect the architecture if they are not met.
3. *Undo a change of subscriber address.* This scenario picks a simple feature that forces you to think about a design of all "undo" functions. This may, in turn, trigger the need to communicate with the end users about what can be undone at reasonable cost.
4. *Complete all the use cases relative to supply-chain management.* The goal of the elaboration phase is also to complete the capture of requirements.

Iteration in the Construction Phase

As the project moves into the construction phase, risks remain a key driver, especially as new and unsuspected risks are uncovered. But the completeness of use cases also becomes a driver. You can plan the iterations feature by feature, trying to complete some of the more critical ones early so that they can be thoroughly tested in more than one iteration. Toward the end of the construction phase, ensuring coverage of full use cases will be the main goal.

The following are examples of iterations in this phase.

1. *Implement all variants of call forwarding, including erroneous ones.* This is a set of related features. One of them may have been implemented during the elaboration phase, and it will serve as a prototype for the rest of the development.
2. *Complete all telephone operator features except night service.* This is another set of features.
3. *Achieve 5,000 transactions per hour on a two-computer setup.* This may increase the required performance relative to what was actually achieved in the previous iteration (only 2,357/hour).

4. *Integrate new version of geographical information system.* This may be a modest architectural change, perhaps necessitated by a problem discovered earlier.
5. *Fix all level 1 and level 2 defects.* This fixes the defects discovered during testing in the preceding iteration that were not fixed immediately.

Iteration in the Transition Phase

The main goal is to finish this generation of the product. Objectives for an iteration are stated in terms of which bugs are fixed and which improvements in performance or usability are included. If you had to drop (or disable) features to get to the end of the construction phase (the IOC milestone, or first beta) on time, you can now complete them or turn them on if they won't jeopardize what has been achieved so far.

The following are examples of iteration goals in this phase.

1. *Fix all severity 1 problems discovered at beta customer sites.* A goal in terms of quality, this may be related to credibility in the market.
2. *Eliminate all start-up crashes caused by mismatched data.* This is another goal expressed in terms of quality.
3. *Achieve 2,000 transactions per minute.* This is performance tuning and involves some optimization: data structure change, caching, and smarter algorithms.
4. *Reduce the number of different dialog boxes by 30%.* Improve usability by reducing visual clutter.
5. *Produce German and Japanese versions.* The beta was produced only for English customers because of the lack of time and to reduce rework on the beta release.

Detail the Work in the Iteration

Now that we have objectively defined the pending iteration's goal, we can proceed to the detailed planning stage as for any other project.

After the scenarios or full-blown use cases to be developed (plus defects to be fixed) have been selected and briefly sketched, you must determine the artifacts that will be affected.

- Which classes are to be revisited?
- Which subsystems are affected or created?
- Which interfaces must probably be modified?
- Which documents must be updated?

The next step is to identify in the Rational Unified Process the activities that are involved and to place them in your project plan. Some activities are done once per iteration (building an iteration plan), whereas others must be done once per class, per use case, or per subsystem (designing a class). Connect the activities with their obvious dependencies and allocate an estimated effort. Most of the activities described in the process are small enough to be accomplished by one person or a small group of people in a matter of a few hours or a few days.

You will probably discover that there isn't enough time in the iteration to accomplish all this work. Rather than extend the iteration (and thereby extend the final delivery time or reduce the number of iterations), reduce the ambitiousness of the iteration. Depending on which phase you are in, make the scenarios simpler or eliminate or disable features.

SUMMARY

- The project management workflow of the Rational Unified Process is useful for balancing competing objectives, managing risk, and overcoming constraints to successfully deliver a product that meets the needs of the customers (the ones who pay the bills) and the end users.
- In an iterative process, the development should be based on a phase plan and a series of iteration plans.
- Risk is a driver for planning.
- Measurement is a key technique used to control projects.
- In building a phase plan, you must assess trade-offs between staff, schedule, and project scope.
- The criteria to define the scope of an iteration vary from phase to phase.

Chapter 8
The Business Modeling Workflow

with Maria Ericsson

THIS CHAPTER EXPLAINS why you might perform business modeling before a system development effort and discusses several ways of doing it. It also briefly describes how to derive software requirements from business models.

PURPOSE

The goals of business modeling are as follows:

- To understand the structure and the dynamics of the organization
- To ensure that customers, end users, and developers have a common understanding of the organization
- To derive the system requirements needed to support the organization

To achieve these goals, the purpose of the business modeling workflow is to develop a model of the business. This model comprises a business use-case model and a business object model.

WHY BUSINESS MODELING?

The majority of software applications are no longer "gizmos" built by programming wizards to be used by computer hobbyists who appreciate technically elegant features for their own sake. Instead, most applications are everyday tools used by people in their working environments as well as at home. These applications need to fit intuitively into the organization in which they are used. We must be sure that the applications we build help people in their daily chores and that they are not perceived as gadgets that do not offer added value. To better fulfill these needs, the new standard is to try to understand the business domain before, or in parallel with, a software engineering project.

Business modeling is not something that we recommend for every software engineering effort. It appears that business models add more value when there are more people directly involved in using the system and more information to be handled by the system. For example, if you were simply adding a new feature to the software of an existing telecommunication switch, you would not consider business modeling. On the other hand, if you were building an order management system to support the sales of telecommunication network solutions, business modeling would be valuable. The sales and order processes in this domain are complex because what you are selling is a custom-made solution and not an off-the-shelf product.

USING SOFTWARE ENGINEERING TECHNIQUES FOR BUSINESS MODELING

One of the major advantages of using a similar modeling technique for business modeling as for software engineering is that you are speaking the same language. It becomes much easier to understand how something described in the business domain might relate to something belonging in the software domain. It also becomes simpler to describe the relationships between artifacts in business models and corresponding artifacts in system models.

Historically, we have seen that modeling techniques developed and matured in the software domain inspire new ways of visualizing an organization. Because object-oriented visual modeling techniques are common for new software projects, using similar techniques in the business domain comes naturally.[1]

DIFFERENT LEVELS OF BUSINESS MODELING

There are different reasons for building business models. Its underlying purpose in the Rational Unified Process is primarily to create a chart of an organization so that we can better understand the requirements of applications used in the organization.

A few years ago, the term *business-process reengineering* (BPR) was very popular and meant a "revolutionary approach to reorganization."[2] Compared with business modeling as we discuss it here, BPR is a much more ambitious undertaking, including charting the existing organization as well as envisioning, building, and implementing a new way of doing things. Our modeling technique can be used as part of this process, too. One difference is that you typically have one or a few business modeling projects that spawn several software engineering projects. Also, there might be significant differences in the number of people issues and political challenges.

WORKERS AND ARTIFACTS

In the Rational Unified Process, how is this translated concretely in terms of workers, artifacts, activities, and workflow? Figure 8-1 shows the workers and artifacts in business modeling.

The main workers involved in business modeling are as follows.

- The *business-process analyst:* The business-process analyst leads and coordinates business use-case modeling by outlining and delimiting the organization being modeled. For example, the business-process analyst establishes which business actors and business use cases exist and how they interact.

1. This idea is developed in the book by Ivar Jacobson, Maria Ericsson, and Agneta Jacobson. *The Object Advantage: Business Process Reengineering with Object Technology*. Reading, MA: Addison-Wesley, 1994.

2. See, for example, the popular book by Michael Hammer and James Champy. *Reengineering the Corporation: A Manifesto for Business Revolution.* New York: HarperBusiness, 1993.

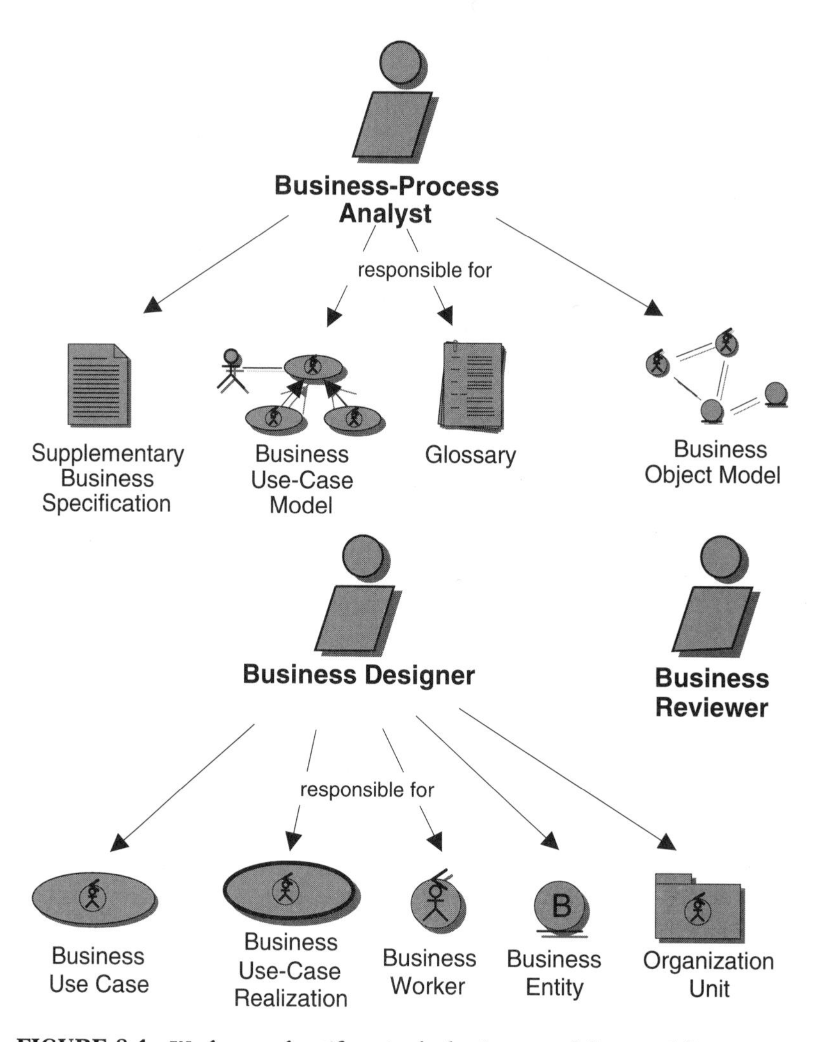

FIGURE 8-1 ***Workers and artifacts in the business modeling workflow***

- The *business designer:* The business designer details the specification of a part of the organization by describing the workflow of one or several business use cases. He or she specifies the business workers and business entities needed to realize a business use case and distributes the behavior of the business use case to these business entities. The business designer defines the responsibilities, operations, attributes, and relationships of one or several business workers and business entities.

Also involved in this workflow are

- *Stakeholders,* who represent various parts of the organization and provide input and review
- The *business reviewer,* who reviews the resulting artifacts

The key artifacts of business modeling are as follows.

- A *business use-case model:* a model of the business's intended functions. It is used as an essential input to identify roles and deliverables in the organization.
- A *business object model:* an object model describing the realization of business use cases.

Other artifacts include the following.

- *Supplementary business specifications:* This document presents any necessary definitions of the business not included in the business use-case model or the business object model.
- A *glossary* defines important terms used in the business.

A business use-case model consists of business actors and business use cases. The actors represent roles external to the business (for example, customers), and the business use cases are processes. A business object model includes business use-case realizations, which show how the business use cases are "performed" in terms of interacting business workers and business entities.

To reflect groups or departments in an organization, business workers and business entities may be grouped into organization units. This parallels the organization of the use-case model and the design model we describe in Chapters 9 and 10. We are using the same modeling technique but at a higher level of abstraction. For example, instead of representing a responsibility in a system, a class at the business level represents a responsibility in an organization.

WORKFLOW

Figure 8-2 shows a typical workflow in business modeling. Working with various business stakeholders, the business-process analyst identifies the key business processes and the business participants involved and describes them in the form of a business use-case

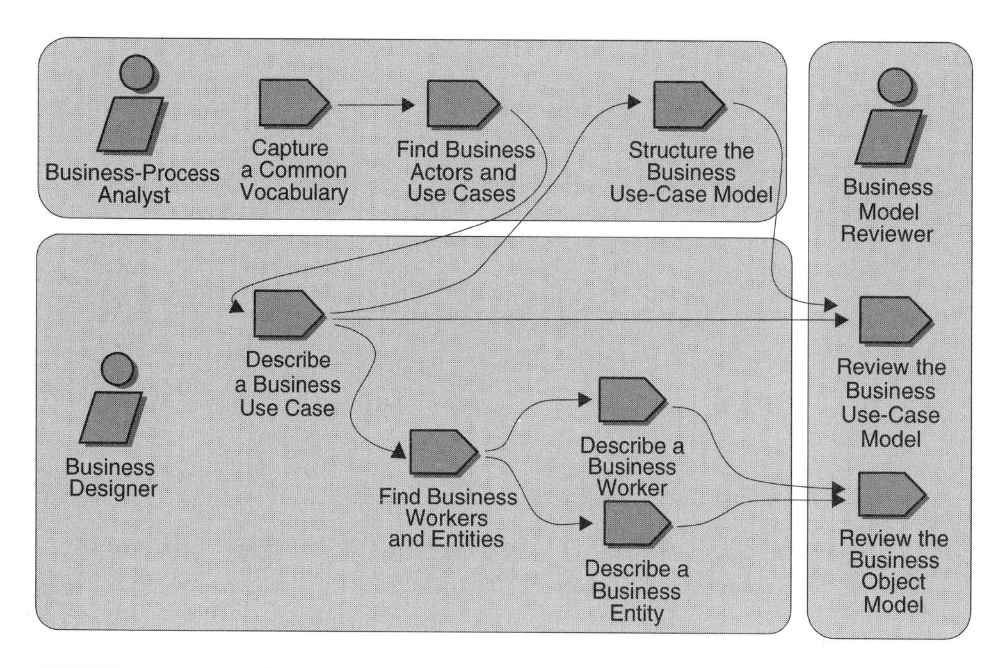

FIGURE 8-2 *A workflow in business modeling*

model composed of business actors and business use cases. During this process, the business-process analyst captures important definitions of terms and concepts in the business.

The business designer details the descriptions of the business use cases, identifies the roles and responsibilities that are needed in the organization, and identifies the deliverables of the processes.

The business reviewer represents the different types of people who are brought in to verify that business models correctly reflect the organization. Each individual covers a specific area of competence. Some people may look only at the higher-level business use-case descriptions, whereas others are better at scrutinizing descriptions of the more detailed business workers and business entities.

FROM THE BUSINESS MODELS TO THE SYSTEMS

One advantage of our approach to business modeling is its clear and concise way of showing dependencies between business and system models. This relationship is shown in Figure 8-3.

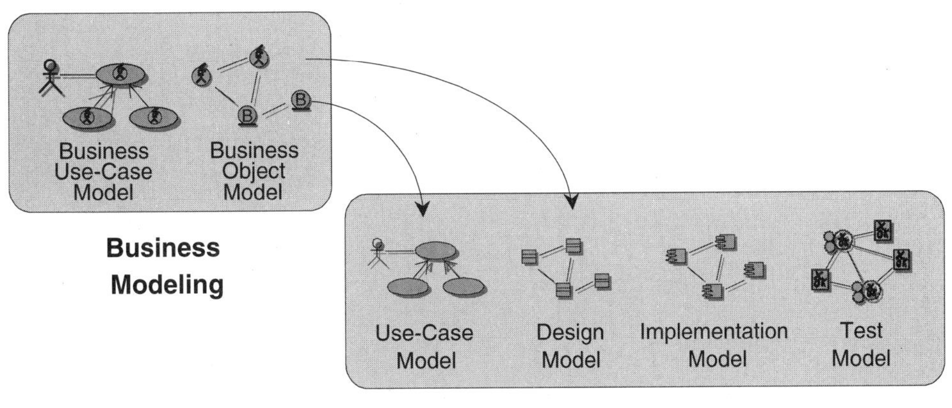

FIGURE 8-3 *From the business models to the system models*

The translation between the two types of models can be summarized as follows.

- Business workers will be played by people who may act as actors to the system.
- Behaviors described for business workers are things that can be automated, so they help us find system use cases.
- Business entities are things we may want the system to help us maintain, so they help us find entity classes in the analysis model of the system.

MODELING THE SOFTWARE DEVELOPMENT BUSINESS

If you were now to revisit Chapter 3, Static Structure: Process Description, you would notice its strong relationship to the concepts introduced in this chapter. That is no accident. The modeling of the Rational Unified Process is based on the same techniques: it is a model of the business of developing software.[3]

3. See Ivar Jacobson and Sten Jacobson, "Reengineering Your Software Engineering Process," *Object Magazine*, March-April 1995, as well as Chapter 9 in *The Object Advantage*, op. cit.

TOOL SUPPORT

Because we are using the same UML concepts (except for slightly different stereotypes) to model a business as we use to model a system, we can use the same tools. Rational Rose provides all we need to produce the visual models described earlier.

Just as in software engineering, you can use the Rational Requisite Pro tool to capture textual aspects of the models and to maintain dependencies between model elements in different models.

To generate and maintain documentation of the models, you can use the Rational SoDA tool for automating documentation.

SUMMARY

- You might need to model a business to understand its structure and dynamics, to ensure that all stakeholders have a common understanding of the organization, and to derive system requirements to support the organization.
- Software engineering techniques can be translated and used for business modeling.
- There are several variants of business modeling, and selecting a suitable one depends on the character of the system you are building and the context of the project.
- You can derive software requirements from models of the business.
- You can use Rational tools to support business modeling.

Chapter 9

The Requirements Workflow

with Maria Ericsson

THIS CHAPTER DEFINES the concepts of requirements and user-interface design and gives a brief overview of the requirements workflow.

PURPOSE

The goals of the requirements workflow are as follows:

- To come to an agreement with the customer and the users on what the system should do
- To give system developers a better understanding of the requirements of the system
- To define the functionality of the system
- To provide a basis for planning the technical contents of iterations
- To provide a basis for estimating cost and time to develop the system
- To define a user interface for the system

To achieve these goals, the requirements workflow describes how to define a vision of the system and translate the vision into a use-case model that, together with supplementary specifications, defines the detailed requirements of the system. In addition, the requirements workflow defines how you manage scope and manage changing requirements by using requirements attributes. Finally, the requirements workflow describes how you define and prototype a user interface.

WHAT IS A REQUIREMENT?

A *requirement* is a condition or capability to which a system must conform.

In his efforts to establish quality criteria as a basis for metrics for evaluating software systems, Robert Grady categorized the necessary quality attributes of a software system as functionality, usability, reliability, performance, and supportability. This model, often referred to as the FURPS model, was used at Hewlett Packard.[1]

Functional Requirements

When we think about the requirements of a system, we naturally tend to think first about those things that the system does on behalf of the user. After all, we developers like to think that we have a "bias for action," and we expect no less of the systems we build. We express these actions as *functional* requirements, which specify the actions that a system must be able to perform.

Functional requirements are used to express the behavior of a system by specifying both the input conditions and output conditions that are expected to result.

Nonfunctional Requirements

To deliver the desired quality to the end user, a system must exhibit a wide variety of attributes that do not specifically relate to the system's functionality. For the system to exhibit these attributes, the implication is that an additional set of requirements must be im-

---

1. Robert Grady, *Practical Software Metrics for Project Management and Process Improvement*. Englewood Cliffs, N.J.: Prentice Hall, 1992, p. 32.

posed. We call this set of requirements *nonfunctional* requirements, and they are every bit as important to the end-user community as are the functional requirements.

The following list summarizes the quality attributes of a system.

- *Usability*
  Usability requirements cover human factors—aesthetics, ease of learning, and ease of use—and consistency in the user interface, user documentation, and training materials.
- *Reliability*
  Reliability requirements cover frequency and severity of failure, recoverability, predictability, and accuracy.
- *Performance*
  Performance requirements impose conditions on functional requirements—for example, a requirement that specifies the transaction rate, speed, availability, accuracy, response time, recovery time, or memory usage with which a given action must be performed.
- *Supportability*
  Supportability requirements cover testability, maintainability, and the other qualities required to keep the system up-to-date after its release. Supportability requirements are unique in that they are not necessarily imposed on the system itself but instead often refer to the process used to create the system or various artifacts of the system development process. An example is the use of a specific C++ coding standard.

Frankly, it's not particularly important to divide requirements into these various categories, and there is little value in a debate over whether a specific requirement is a usability or a supportability requirement. The value in these definitions is that they provide a template for requirements elicitation and for assessing the completeness of your understanding. In other words, if you ask about and come to understand requirements of all these categories, you have a much higher degree of certainty that you have truly understood the critical requirements before you begin substantial investment in development. Often you will find, for example, that certain reliability requirements are implied by given functional requirements, and it is equally important to explore these reliability requirements. A system that fails to meet an implied reliability or performance requirement fails just as badly as one that fails to meet an explicit functional need.

DIFFERENT TYPES OF REQUIREMENTS

Traditionally, requirements are seen as detailed specifications that fit into one of the categories just mentioned and to which the system must conform, expressed in the form "The system shall . . ." To effectively manage the full requirements process, however, it is helpful to gain a more comprehensive understanding of the actual *needs* that are to be fulfilled by the system being developed. This understanding arms the development team with the *why* ("why does this system need to operate at 99.3% accuracy?") as well as the *what*. Armed with this understanding, the team will be able to do a better job of interpreting the requirements ("Does it also need to operate at 99.3% accuracy during the time when routine maintenance is happening?") as well as being able to make trade-offs and design decisions that optimize the development process ("If we can achieve 92% accuracy with half the effort, is that a reasonable trade-off?").

Stakeholder Needs

It seems obvious that the development team must understand the specific needs of the users of the system as well as the other stakeholders. A stakeholder is any person or representative of an organization who has a stake—a vested interest—in the outcome of a project. A stakeholder can be an end user, a purchaser, a contractor, a developer, a project manager, or anyone else who cares enough or whose needs must be met by the project. For example, a nonuser stakeholder might be a system administrator whose workload will depend on certain default conditions imposed by the system software on the user. Other nonuser stakeholders might represent the economic beneficiary of the system ("We could have eliminated that user feature on the cell phone had we known it would increase our roaming charges"). This set of requirements, which we call *stakeholder needs*, provides a crucial piece of the puzzle that allows us to determine both the *why*s and the *what*s of system behavior. Often, these needs will be vague and ambiguous ("I need easier ways to share my knowledge of project status" or "I need to increase my personal productivity"). Yet they set a most important context for all that follows.

System Features

Interestingly, when you initially interview stakeholders about their needs and requirements, they typically describe neither of the entities just defined. Typically, they tell you neither their real need ("If Joe and I don't start communicating better, our jobs are at risk" or "I want to be able to slow this vehicle down as quickly as possible without skidding") nor the actual requirement for the system ("I must be able to attach an RTF file to an e-mail message" or "The vehicle shall have a computer control system dedicated to each wheel that . . ."). Instead, they describe an abstraction of both ("I want automatic e-mail notification" or "I want a vehicle with ABS").

We call this type of abstraction—these higher-level expressions of system behavior—the *features* of the system. Technically, we can think of a feature as a service to be provided by the system that directly fulfills a user need. These features are often not very well defined, and they may even be in conflict with one another ("I want ABS" and "I want lower maintenance requirements"), but nonetheless they are a representation of real needs. What is happening in this discussion is that the stakeholder has already translated the real need ("better communication with Joe") into a behavior of the system ("automatic e-mail notification"). In so doing, there has been a subtle shift in the stakeholder's thinking from the *what* ("better communication") to the *how*—that is, how it will be implemented ("automated e-mail notification").

Because it is easy to discuss these features in natural language and to document and communicate them, they add important context to the requirements workflow. In the Rational Unified Process, we simply treat these features as another type of requirement. In addition, by adding to these features various attributes—such as risk, level of effort, and customer priority—you add a richness to your understanding of the system. You can then use this enhanced understanding to effectively manage scope before substantial investments have been made in low-priority features.

Software Requirements

Neither an understanding of stakeholder needs nor an understanding of system features is, by itself, adequate to communicate to developers exactly what the system should do ("Hey, Bill, go code an automatic e-mail notification system"). You need an additional level of specificity that translates these needs and features into specifications that you can design, implement, and test to determine system conformance. At this level of specificity, you must deal with the functional and nonfunctional requirements of the system.

Specifying Requirements with Use Cases

In Chapter 6, A Use-Case-Driven Process, we learned about use-case modeling, a powerful technique that we can use to express the detailed behavior of the system via simple threads of user-system interactions that can be easily understood by users as well as developers. A use-case model is a model of the system and its intended behavior. It consists of a set of actors to represent external entities that interact with the system along with use cases that define how the system is used by the actors. The use-case model is an effective way of expressing these more detailed software requirements.

CAPTURING THE REQUIREMENTS

For each project, you can now see a common structure among these expressions of requirements, as depicted in Figure 9-1. First, you gather the collection of stakeholder needs. They encompass all the requests and wish lists you get from end users, customers, marketing people, and other project stakeholders.

The set of stakeholder needs is used to develop a *vision document* containing the set of high-level features of the system. These system features express services that must be delivered by the system in order to fulfill the stakeholder needs. We call this relationship *requirements traceability*. The features you decide to include in the vision document are based on an analysis of the cost of implementing a desired feature and a determination of the return on that investment. This analysis is found in the business case for the project.

Before coding begins, you must translate these features into detailed software requirements at a level at which you can design and build an actual system and identify test cases to test the system behavior. These detailed requirements are captured in two places: in the use-case model and in supplementary specifications. As you are detailing the requirements, you must keep in mind the original stakeholder needs and the system features to ensure that you are interpreting the vision correctly. This constant consideration of fea-

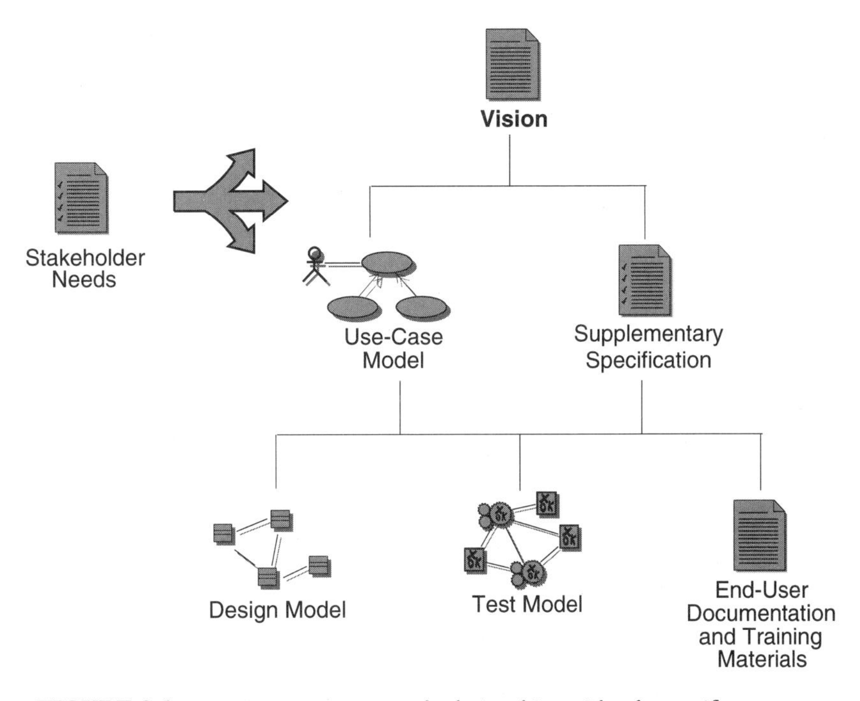

FIGURE 9-1 ***Requirements types and relationships with other artifacts***

tures and software requirements implies a traceability relationship wherein software requirements are derived from one or more features of the system, which in turn fulfill one or more stakeholder needs.

The detailed requirements are then realized in a design model and end-user documentation and are verified by a test model.

The vision and stakeholder needs documents are completed early in the project. For the remainder of the requirement structure, you will, based on risk and architectural stability, divide them into "slices" and detail them in an iterative fashion. As you are detailing these requirements, you will find flaws and inconsistencies that make it necessary to go back to the stakeholders for clarifications and trade-offs and to update the stakeholder needs and vision.

DESIGNING THE USER INTERFACE

The expression *user-interface design* can mean one of at least the two following things:

- The visual shaping of the user interface so that it handles various usability requirements
- The design of the user interface in terms of design classes (and components such as ActiveX classes and JavaBeans) that is related to other design classes dealing with business logic, persistence, and so on, and that leads to the final implementation of the user interface

In the requirements workflow, we operate under the first definition. The main input is the use-case model and the supplementary specification, or at least outlines thereof. The results are detailed definitions of user characteristics and realizations of the user-interface-specific parts of the use cases.

Based on these results, you build a prototype of the user interface, in most cases by using a prototyping tool. We call this *user-interface prototyping.*

WORKERS AND ARTIFACTS

In the Rational Unified Process, how is all this translated concretely in terms of workers, artifacts, activities, and workflow? Figure 9-2 shows the workers and artifacts in the requirements workflow.

The main workers involved in the requirements workflow are as follows.

- *System analyst*
  The system analyst leads and coordinates requirements elicitation and use-case modeling by outlining the system's functionality and delimiting the system.
- *Use-case specifier*
  The use-case specifier details all or part of the system's functionality by describing the requirements aspect of one or several use cases.
- *Architect*
  In the requirements workflow, the architect is responsible for identifying the architecturally significant use cases and requirements and contributing to their definition.

The user interface is then more precisely defined by the user-interface designer, who leads and coordinates the prototyping and

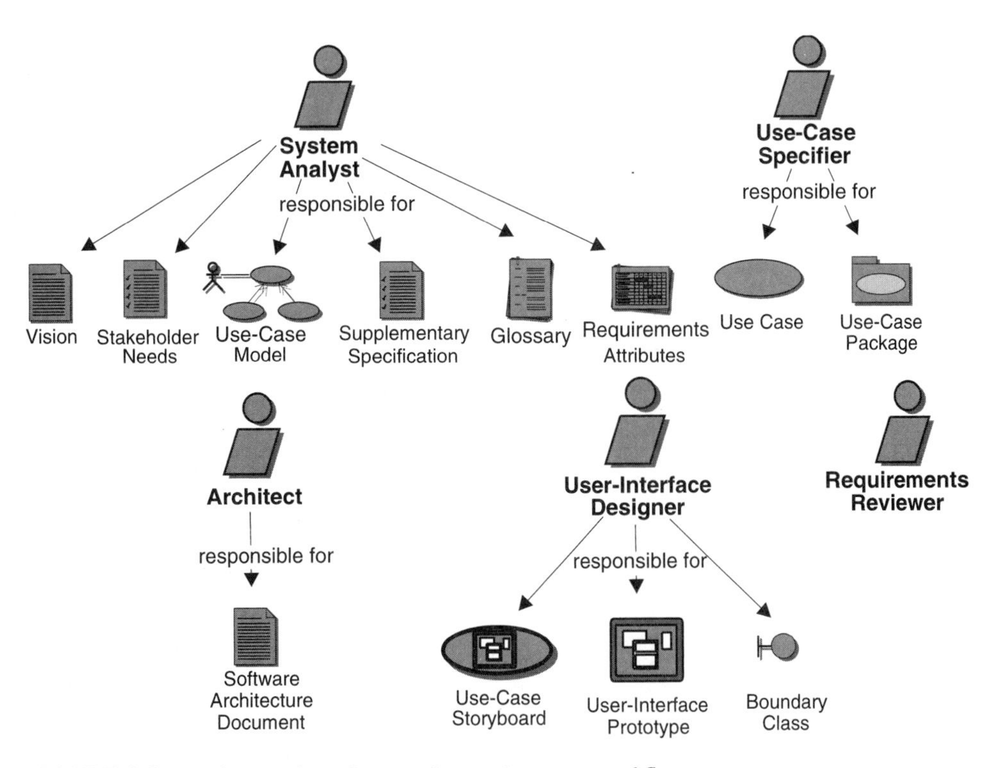

FIGURE 9-2 *Workers and artifacts in the requirements workflow*

design of the user interface. The requirements reviewer is involved in reviewing the resulting artifacts.

This workflow produces several artifacts. High-level artifacts are used to manage the project itself as well as its scope:

- Stakeholder needs
- Vision document

It also includes more detailed artifacts such as

- Use-case model
- Supplementary specifications (covering detailed requirements that do not easily fit into the use-case format)
- Requirements attributes (which define and represent the characteristics of the various types of requirements as well as dependencies between them)

The use-case model is expressed in terms of : (1) actors, (2) use cases, and (3) use-case packages. Finally, the following artifacts define the user interface:

- Use-case storyboard
- User-interface prototype
- Boundary classes (to show how the user interface is realized in the analysis model)

WORKFLOW

Figure 9-3 summarizes the activities that constitute the requirements workflow. Working with stakeholders of the project, the system analyst understands what the system must do and perhaps what it should not do and also identifies nonfunctional requirements. The

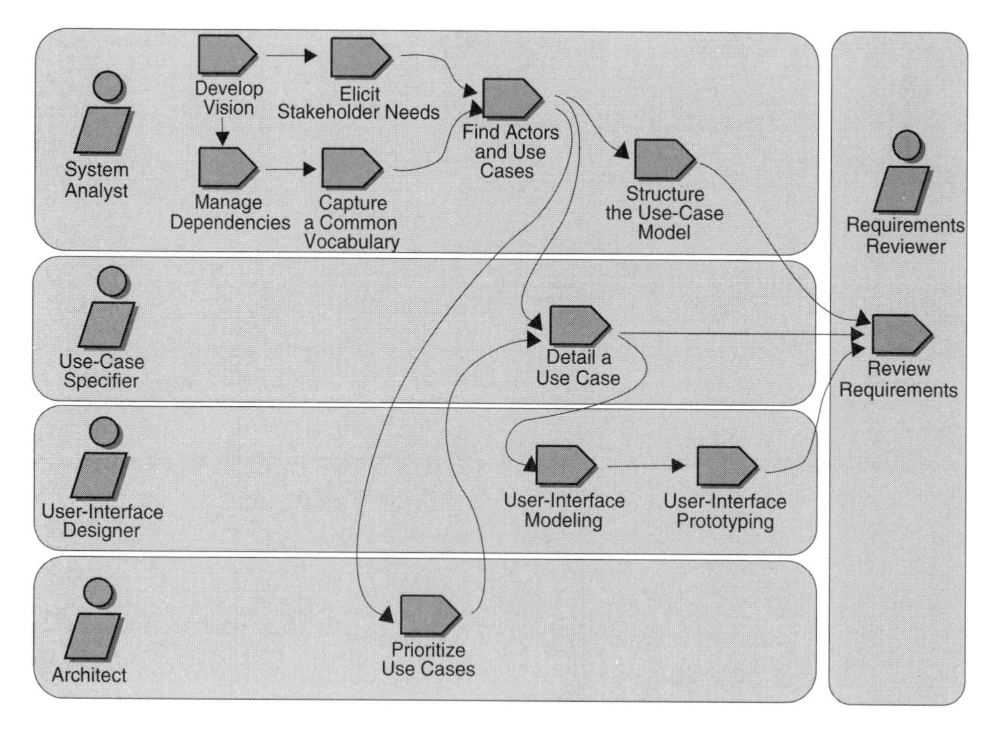

FIGURE 9-3 *Workflow in requirements*

system analyst can then develop a vision for the project. This vision, expressed as a set of features written from the stakeholder's perspective, is a basis for developing an outline of a use-case model.

The use-case specifier is assigned a set of use cases and supplementary requirements, which he or she will detail and make consistent with other requirements workflow artifacts. The use-case specifier does not work in isolation but rather should communicate regularly with other individual use-case specifiers as well as with system analysts.

The user-interface designer works in parallel with the use-case specifier to design the system's user interface. In most cases, there is a synergistic interaction between the two.

The architect is involved primarily in earlier iterations and works with the system analyst and use-case specifier to ensure the integrity of the architecturally significant use cases.

The requirements reviewer represents all the different kinds of people you bring in to verify that the requirements are correctly perceived and interpreted by the development team. Because reviews have differing purposes, they are performed several times during the execution of the requirements workflow by requirements reviewers with different profiles.

Together, the individuals acting as these workers form a team that exercises the following skills.

- *Analyze the problem:* Taking nothing for granted, pursue stakeholders to understand the problem behind the problem.
- *Elicit stakeholder needs:* Adopt various elicitation techniques to obtain a detailed understanding of the needs to be met by the system.
- *Define the system:* Establish the boundaries of what the system should do and set the right expectations among the stakeholders.
- *Manage the scope of the project:* Along the way, continuously monitor the scope and progress of the project to ensure that schedules are met in a controlled and expected fashion.
- *Detail the system:* Use effective techniques to define the details of what the system is to be capable of.

- *Manage changing requirements:* Be prepared for and capable of handling changes in requirements that will occur as the project unfolds.

TOOL SUPPORT

Rational Requisite Pro supports you in capturing requirements, organizing them in documents and in the database, and managing requirement scope and change. Moreover, if you are using use cases, Requisite Pro will help you in describing the textual properties of the use cases.

For visual modeling of requirements artifacts (use-case model, use-case storyboards, boundary classes), you can use Rational Rose. Having requirements artifacts in Rose allows you to maintain dependencies to elements in the design model.

Rational SoDA helps automate the generation of documentation. It allows you to define an "intelligent template" that can extract information from various sources. Rational SoDA is particularly useful if you are using several tools to capture the results of your workflow, but you must produce final documentation that pulls together that information in one place.

SUMMARY

- A requirement is a condition or capability to which a system must conform.
- In a typical project, you must consider several types of requirements, including functional and nonfunctional requirements.
- The Rational Unified Process defines user-interface design as the visual shaping of the user interface so that it handles various usability requirements.
- In the basic requirements workflow, various workers analyze the problem, elicit stakeholder needs, define the system, manage the scope of the project, detail the system, and manage changing requirements.
- Rational tools support the capture and management of requirements.

Chapter 10
The Analysis and Design Workflow

with Kurt Bittner

THIS CHAPTER DESCRIBES the key aspects of the analysis and design workflow. It introduces the main artifacts used in design—classes, subsystems, and collaborations—and explains the role of analysis and design in the overall development effort.

PURPOSE

The purpose of the analysis and design workflow is to translate the requirements into a specification that describes how to implement the system. To make this translation, you must understand the requirements and transform them into a system design by selecting the best implementation strategy. Early in the project, you must establish a robust architecture so that you can design a system that is easy to understand, build, and evolve. Then you must adjust the design to match the implementation environment, designing it for performance, robustness, scalability, and testability, among other qualities.

ANALYSIS VERSUS DESIGN

The purpose of analysis is to transform the requirements of the system into a form that maps well to the area of concern of the software designer—that is, to a set of classes and subsystems. This transformation is driven by the use cases and is further shaped by the system's nonfunctional requirements. Analysis focuses on ensuring that the system's functional requirements are handled. For simplicity's sake, it ignores many of the nonfunctional requirements of the system and also the constraints of the implementation environment. As a result, analysis expresses a nearly ideal picture of the system.

The purpose of design, on the other hand, is to adapt the results of analysis to the constraints imposed by nonfunctional requirements, the implementation environment, performance requirements, and so forth. Design is a refinement of analysis. It focuses on optimizing the system's design while ensuring complete requirements coverage.

HOW FAR MUST DESIGN GO?

Design must define only enough of the system so that it can be implemented unambiguously. What constitutes "enough" varies from project to project and company to company. In some cases, the design may be thoroughly elaborated to the point that the system can be implemented through a direct and systematic transformation of the design into code.

In other cases, the design more closely resembles a sketch, elaborated only far enough to ensure that the implementer can produce a set of components that satisfy the requirements. The degree of specification therefore varies with the expertise of the implementer, the complexity of the design, and the risk that the design might be misconstrued.

When the design is specified precisely, the code can be closely tied to the design and can be kept synchronized with the design in what we call *round-trip engineering*, thereby avoiding one transformation step and a potential source of error.

When the degree of completeness and precision in the design is very high—so high that you can execute the transformation directly by interpreting the design or by rapidly generating from it small amounts of code—the transformation is almost invisible to the designer, and the design appears to be "executable."

WORKERS AND ARTIFACTS

The Rational Unified Process expresses the analysis and design process in terms of workers, artifacts, activities, and workflow. Figure 10-1 shows the workers and artifacts involved.

The main workers involved in analysis and design are as follows.

- *Architect*
  The architect leads and coordinates technical activities and artifacts throughout the project. He or she establishes the overall structure for each architectural view: the decomposition of the view, the grouping of elements, and the interfaces between the major groupings. In contrast with the views of the other workers, the architect's view is one of breadth rather than depth. See Chapter 5, An Architecture-centric Process.
- *Designer*
  The designer defines the responsibilities, operations, attributes, and relationships of one or several classes and determines how they should be adjusted to the implementation environment. In addition, the designer may have respon-

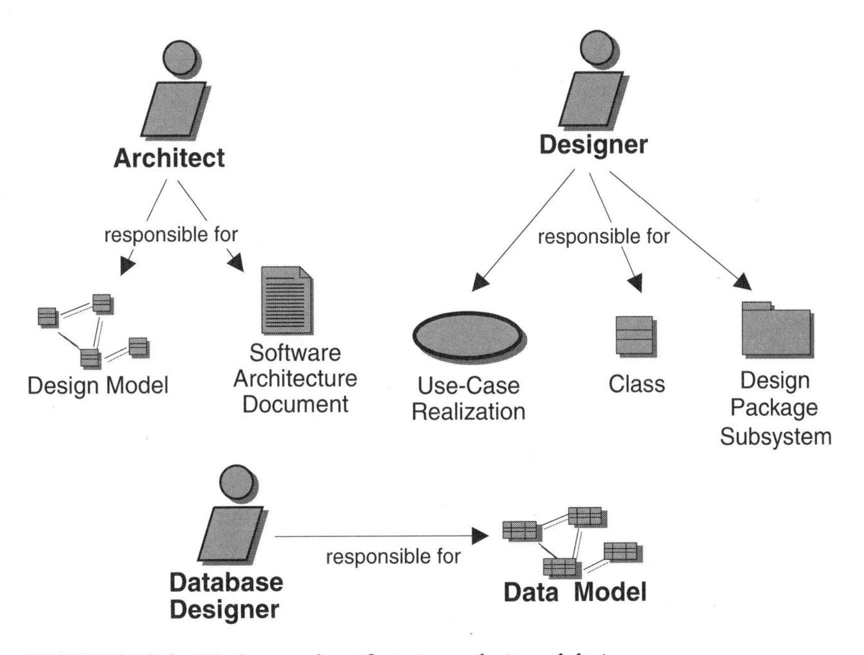

FIGURE 10-1 *Workers and artifacts in analysis and design*

sibility for one or more design packages or design subsystems, including any classes owned by the packages or subsystems.

Analysis and design can optionally include the following workers.

- *Database designer:* The database designer is needed when the system being designed includes a database.
- *Architecture reviewer and design reviewer:* These specialists review the key artifacts produced through this workflow.

The key artifacts of analysis and design are as follows:

- The *design model,* which is the major blueprint for the system under construction
- The *software architecture document,* which captures various architectural views of the system

THE DESIGN MODEL

The primary artifact of the analysis and design workflow is the design model. It consists of a set of collaborations of model elements that provide the behavior of the system. This behavior in turn is derived primarily from the use-case model and from nonfunctional requirements.

The design model consists of collaborations of classes, which may be aggregated into packages and subsystems to help organize the model and to provide compositional building blocks within the model. A *class* is a description of a set of objects that share the same responsibilities, relationships, operations, attributes, and semantics. A *package* is a logical grouping of classes, perhaps for organizational purposes, that reduces the complexity of the system. A *subsystem* is a kind of package consisting of a set of classes that act as a single unit to provide specific behaviors.

THE ANALYSIS MODEL

Generally, there is one design model of the system; analysis produces a rough sketch of the system, which is further refined in design. The upper layers of this model describe the application-specific, or more analysis-oriented, aspects of the system. Using a single model reduces the number of artifacts that must be maintained in a consistent state.

In some companies—those in which systems live for decades or there are many variants of the system—a separate analysis model has proven useful. The analysis model is an abstraction, or generalization, of the design. It omits most of the details of how the system works and provides an overview of the system's functionality. The extra work required to ensure that the analysis and design models remain consistent must be balanced against the benefit of having a view of the system that represents only the most important details of how the system works.

THE ROLE OF INTERFACES

Interfaces are used to specify the behavior offered by classes, subsystems, and components in a way that is independent of the implementation of the behavior. They specify a set of operations performed by the model elements, including the type returned and the number and types of parameters. Any two model elements offering the same interface are interchangeable with each other. Interfaces improve the flexibility of designs by reducing dependencies between different parts of the system and therefore making them easier to change.

COMPONENT-BASED DESIGN

Components are model elements in the implementation model and are used to express the way the system is implemented in a set of smaller "chunks." Components offer one or more interfaces and are dependent only on the interfaces of other components. Using components lets you design, implement, deliver, and replace various parts of the system independently of the rest of the system.

The design model element that corresponds to the component is the subsystem: subsystems are independent units of functionality that allow different parts of the system to be designed and implemented independently. Subsystems offer interfaces and are dependent on the interfaces of other model elements.

If necessary, the design can also include a data model (expressed in UML) that expresses the way the persistent objects in the system are stored and retrieved. A data model is useful when the underlying database technology imposes a significantly different physical model from that of the object-oriented application.

WORKFLOW

Figure 10-2 illustrates a typical workflow in analysis and design. Two workflow details are interwoven in analysis and design: one is centered on the architecture, and the other one is centered on a more detailed design of classes and subsystems.

Let's look first at the architecture level.

- In architectural analysis, the architect defines how the architecture will be organized. At the top level, the architect identifies the fundamental architectural patterns, key mechanisms, and modeling conventions for the system. For

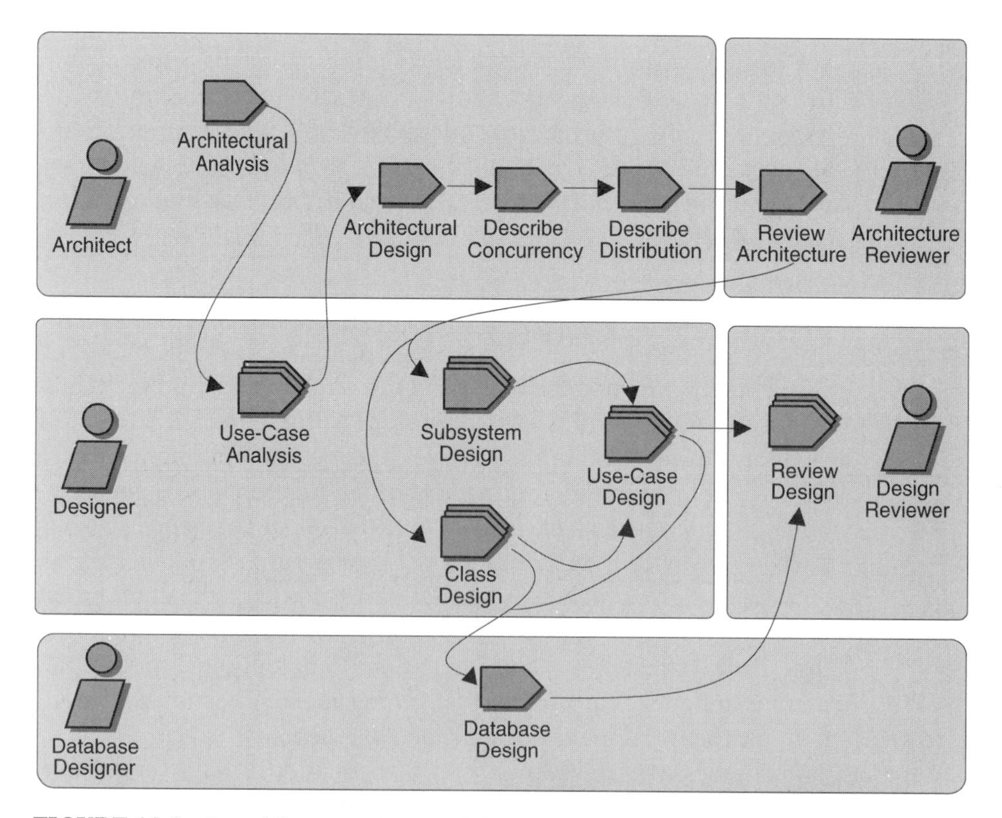

FIGURE 10-2 *A workflow in analysis and design*

example, this worker defines the layering, the principles for organizing subsystems, and the reuse strategy. This analysis provides vital input to the planning process.

- After enough use-case analysis has been done, the resulting classes are used to initiate architectural design: architecturally significant classes are identified and grouped into design packages and subsystems. Then the subsystems are organized in layers.
- Design of the architecture continues with the other architectural views as needed. "Describe Concurrency" creates and maintains the process view; "Describe Distribution" creates and maintains the deployment view. The creation and maintenance of the implementation view are covered in Chapter 11, The Implementation Workflow.

Now let's look at the class and subsystem design level.

- Starting with the results of use-case analysis, in which use-case realizations were sketched, and using the architecture, the designer identifies class characteristics and relationships and further refines the classes.
- Subsystem interfaces are defined. The behavior offered by the subsystem through its interfaces is further refined in terms of the contained classes, which realize the interface operations.
- The use cases can then be fully designed, with each use-case realization completely specified in terms of operations on classes or subsystems, in sequence diagrams, or collaboration diagrams.

When the system involves a large amount of data in a database, the database designer maps persistent classes to database tables and, if necessary, identifies those aspects of the behavior of persistent classes that are implemented in the database.

This presentation of the activities may make them seem sequential, but remember two things. First, there are many dynamic interactions between the various workers; discovery of a new class or entity may require adjustment of the architecture. And in an iterative process, at each iteration and especially during the elaboration phase, the architecture is still fluid and the more detailed design is still rather "skimpy."

All these activities are revisited again and again, both within an iteration and from iteration to iteration.

TOOL SUPPORT

The language of choice to express all these models is the UML, and the modeling guidelines associated with the various artifacts are expressed in terms of the UML. The tool of choice to capture, manage, and display the models is Rational Rose. Rose allows you to perform round-trip engineering with a few selected programming languages, keeping the design and the code perfectly synchronized and allowing the system to evolve from the design, from the code, or from both.

The Rational Unified Process provides tool mentors to guide the designers in the use of UML and Rose. ObjecTime Developer allows the direct execution of a design model. SoDA allows the automatic creation of documents and reports, extracting and formatting information extracted from several other tools, such as Rose or Requisite Pro.

SUMMARY

- Analysis and design bridge the gap between requirements and implementation. This workflow uses use cases to identify a set of objects that is subsequently refined into classes, subsystems, and packages.
- Responsibilities in analysis and design are distributed among the architect (the big picture issues), the designer (the details), and the database designer (the details that require specialized knowledge about handling persistence).
- Analysis and design result in a design model, which can be abstracted using three architectural views. The logical view presents the decomposition of the system into a set of logical elements (classes, subsystems, packages, and collaborations). The process view maps those elements to the processes and threads in the systems. The deployment view maps the processes to a set of nodes on which they execute.
- In some cases, a separate analysis model can be useful for presenting an overview or abstraction of the system.

Chapter 11
The Implementation Workflow

THIS CHAPTER DESCRIBES THE implementation workflow. It introduces the concepts of prototypes and incremental integration.

PURPOSE

The implementation workflow has four purposes:

- To define the organization of the code in terms of implementation subsystems organized in layers
- To implement classes and objects in terms of components (source files, binaries, executables, and others)
- To test the developed components as units
- To integrate into an executable system the results produced by individual implementers or teams

The scope of the implementation workflow is limited to unit test of individual components. System test and integration test are described in the test workflow (see Chapter 12). This is because the implementer is responsible for unit test. Remember that, in practice, the workflows are interwoven.

To explain implementation in the Rational Unified Process, we introduce the following three key concepts:

- Builds
- Integration
- Prototypes

BUILDS

A *build* is an operational version of a system or part of a system that demonstrates a subset of the capabilities provided in the final product.

Builds are an integral part of the iterative life cycle. They represent ongoing attempts to demonstrate the functionality developed to date. Each build is placed under configuration control in case there is a need to roll back to an earlier version when added functionality causes breakage or otherwise compromises build integrity.

During iterative software development there will be numerous builds. Each build provides early review points and helps to uncover integration problems as soon as they are introduced.

INTEGRATION

The term *integration* refers to a software development activity in which separate software components are combined into a whole. Integration is done at several levels and stages of the implementation:

- To integrate the work of a team working in the same implementation subsystem before the subsystem is released to system integrators
- To integrate subsystems into a complete system

The Rational Unified Process approach to integration is that the software is integrated incrementally. *Incremental* integration means that code is written and tested in small pieces and is then combined into a working whole by the addition of one piece at a time.

The contrasting approach is phased integration. *Phased* integration relies on integrating multiple (new and changed) components at a time. The major drawback of phased integration is that it introduces multiple variables and makes it harder to locate errors. An error could be inside any one of the new components, in the interaction between the new components and those at the core of the system, or in the interaction between the new components.

Incremental integration offers a number of benefits.

- Faults are easy to locate. When a new problem occurs during incremental integration, the new or changed component (or its interaction with the previously integrated components) is the obvious first place to look for the fault. Incremental integration also makes it more likely that defects will be discovered one at a time, something that makes it easier to identify faults.
- The components are tested more fully. Components are integrated and tested as they are developed. This means that the components are exercised more often than they are when integration is done in one step.
- Some part of the system is running earlier. It is better for the morale of developers to see early results of their work instead of waiting until the end to test everything. It also makes it possible to receive early feedback on design, tools, rules, or style.

It is important to understand that integration occurs (at least once) within each iteration. An iteration plan defines the use cases to be designed and thus the classes to be implemented. The focus of the integration strategy is to determine the order in which classes are implemented and then combined.

PROTOTYPES

Prototypes are used in a directed way to reduce risk. Prototypes can reduce uncertainty surrounding the following issues:

- The business viability of a product being developed
- The stability or performance of key technology
- Project commitment or funding (through the building of a small proof-of-concept prototype)
- The understanding of requirements
- The look and feel, and ultimately the usability, of the product

A prototype can help to build support for the product by showing something concrete and executable to users, customers, and managers.

The nature and goal of the prototype, however, must remain clear throughout its lifetime. If you don't intend to evolve the prototype into the real product, don't suddenly assume that because the prototype works it should become the final product. An exploratory, behavioral prototype, intended to rapidly try out a user interface, rarely evolves into a strong, resilient product.

Types of Prototypes

You can view prototypes in two ways: what they explore and how they evolve (their outcome). In the context of the first view—what they explore—there are two main kinds of prototypes:

- A behavioral prototype, which focuses on exploring specific behavior of the system
- A structural prototype, which explores architectural or technological concerns

In the context of the second view—their outcome—there are also two kinds of prototypes:

- An exploratory prototype, also called a throwaway prototype, which is thrown away after it is finished and you have learned whatever you wanted from it
- An evolutionary prototype, which gradually evolves to become the final system

Exploratory Prototypes

An exploratory prototype is designed to be like a small experiment to test a key assumption involving functionality or technology or both. This prototype might be something as small as a few hundred lines of code created to test the performance of a key software or hardware component. Or to clarify requirements, a small prototype might be developed to see whether the developer understands a particular behavioral or technical requirement.

Exploratory prototypes are usually intended to be throwaway prototypes created with minimal effort, and therefore they are usually tested only informally. The design of exploratory prototypes tends to be informal and is usually the work of one or two developers.

Evolutionary Prototypes

Evolutionary prototypes, as their name implies, evolve from one iteration to the next. Although initially they are not production quality, their code tends to be reworked as the product evolves. To keep rework manageable, evolutionary prototypes tend to be more formally designed and somewhat formally tested even in the early stages. As the product evolves, testing as well as design usually becomes more formalized.

Squeezed by time constraints, project teams often fall into the trap of keeping an exploratory or throwaway prototype, with all its shortcomings in quality, as the final system. To avoid this problem, you should clearly define the objective and scope of a prototype when starting to develop it.

Behavioral Prototypes

Behavioral prototypes tend to be exploratory prototypes; they do not try to reproduce the architecture of the system to be developed but instead focus on what the system will do as seen by the users (the "skin"). Frequently, this kind of prototype is "quick and dirty" and not built to project standards. For example, you might use Visual Basic as the prototyping language even though you intend to use C++ for the development project.

Structural Prototypes

Structural prototypes tend to be evolutionary prototypes; they are more likely to use the infrastructure of the ultimate system (the "bones") and are more likely to evolve into the final system. If you build the prototype using the production language and tool set, you gain the added advantages of being able to test the development environment and helping personnel familiarize themselves with new tools and procedures.

The Rational Unified Process advocates the development of an evolutionary structural prototype throughout the elaboration phase, accompanied by any number of exploratory prototypes. An example of an exploratory prototype is the user-interface prototype introduced in Chapter 9.

WORKERS AND ARTIFACTS

In the Rational Unified Process, how is all this translated in terms of workers, artifacts, activities, and workflows? Figure 11-1 shows the workers and artifacts involved in the implementation workflow.

The main workers involved in the implementation workflow are as follows:

- The *implementer,* who develops the components and related artifacts and performs unit testing
- The *system integrator,* who constructs a build

Other workers include

- The *architect,* who defines the structure of the implementation model (layering and subsystems)
- The *code reviewer,* who inspects the code for quality and conformance to the project standard

The key artifacts of implementation are as follows.

- *Implementation subsystems*
  A collection of components and other implementation subsystems. It is used to structure the implementation model by dividing it into smaller parts.
- *Component*
  A piece of software code (source, binary, or executable) or a file containing information (for example, a start-up file or a

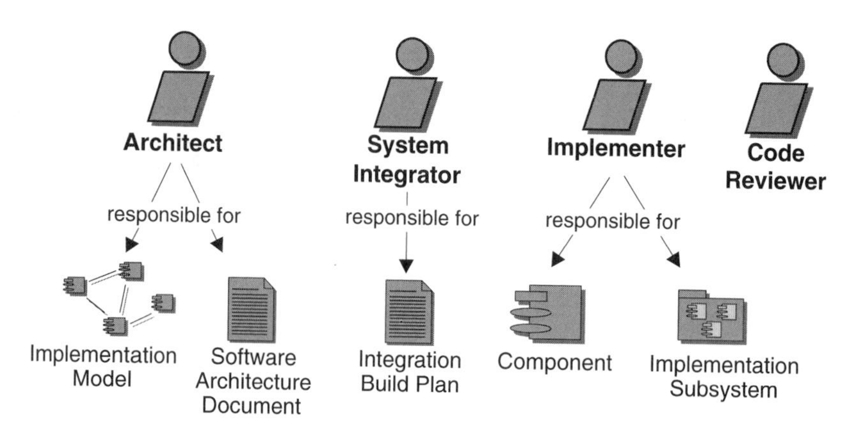

FIGURE 11-1 *Workers and artifacts in implementation*

readme file). A component can also be an aggregate of other components—for example, an application consisting of several executables.

- *The integration build plan*
  This document defines the order in which the components and subsystems should be implemented and specifies the builds to be created when the system is integrated.

WORKFLOW

Figure 11-2 shows an implementation workflow, giving an overview of all activities. Based on the implementation model's structure, the system integrator defines how the system will be integrated.

The implementer in charge of a subsystem or component does the same thing at the level of one subsystem and then proceeds with the actual implementation: developing code and correlated artifacts

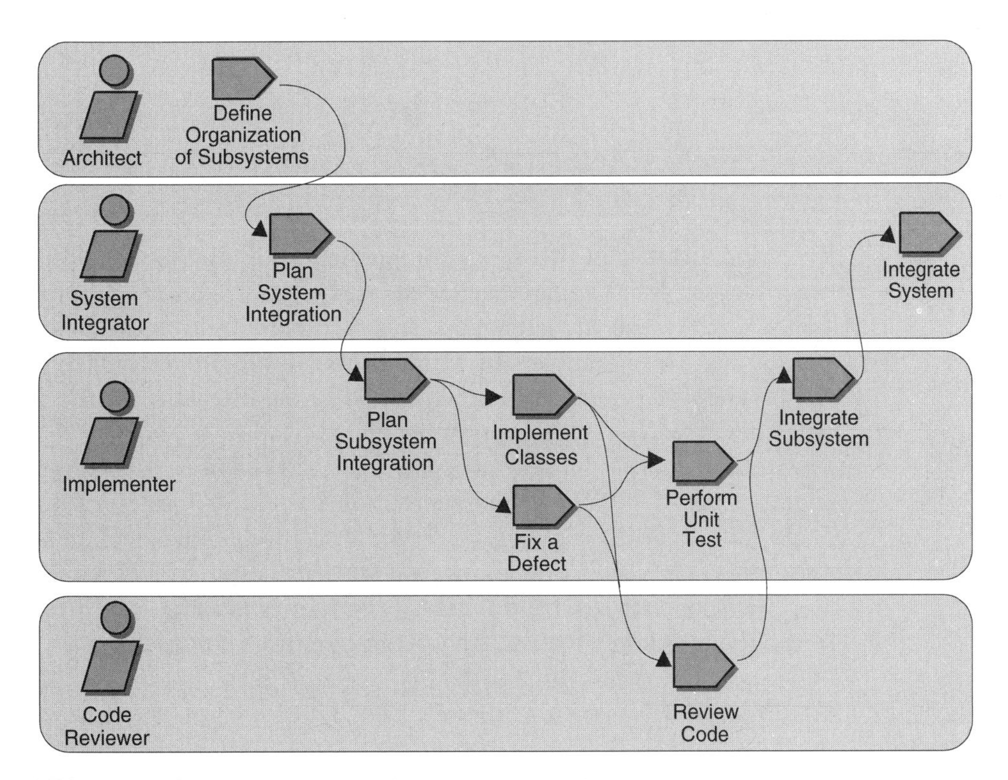

FIGURE 11-2 ***A workflow in implementation***

or, if required, fixing code. The implementer then performs tests at the level of the unit and proceeds to integrate the entire subsystem. The system integrator then proceeds to integrate subsystems together.

Implementation is closely tied to design, and there are clear tracing links from design elements to implementation elements (for example, classes to code). Under certain conditions (such as the use of a specific programming language or type of application), round-trip engineering allows a close tie between design and implementation. The person acting alternately as the designer and the implementer can either modify the design model and regenerate the corresponding code or can modify the implementation code and then alter the design to match the modification. This closes a gap in the process and helps avoid errors in translating the design or a design getting out of sync with the implementation and therefore not being trusted by the implementers.

For information about the integration test and system tests, see Chapter 12, The Test Workflow. Chapter 13, The Configuration and Change Management Workflow, discusses the activity of building the product.

TOOL SUPPORT

Traditionally, it is in the area of implementation that classic software development tools have been used—tools such as editors, compilers, linkers, and debuggers, among others. Today these code-level tools are assembled into integrated development environments and share semantic knowledge. An example is the Rational Apex development environment for Ada and C++.

Rose provides support for round-trip engineering, closely tying together design and the implementation activities. Tools such as Purify and Quantify support the detection of defects in code. ClearCase provides support for individual workspaces as well as "staging" workspaces for subsystem and system integration. ClearQuest is the tool of choice for tracking change requests and defects and tracing them to the source code.

SUMMARY

- A key feature of the Rational Unified Process is its approach to incremental integration throughout the life cycle.
- During the integration phase, you build an evolutionary, structural prototype that becomes the final system.
- In parallel, you can build a few throwaway behavioral prototypes to explore issues such as the user interface.
- Round-trip engineering is a technique, supported by a tool such as Rose, that closely ties together the design and the corresponding implementation of a software development effort.

Chapter 12
The Test Workflow

with Bruce Katz

IN THIS CHAPTER, we explain the concept of quality, describe the test workflow, and then discuss the relationship between quality, test, and the other workflows in the process.

PURPOSE

The purpose of testing is to assess and report the findings regarding the level of quality achieved in the product. The test workflow involves the following:

- Verifying the interaction between objects and components
- Verifying the proper integration of all components of the software
- Verifying that all requirements have been correctly implemented
- Identifying and ensuring that all discovered defects are addressed before the software is deployed

QUALITY

Quality is something we all strive for in our products, processes, and services. Yet when people are asked to define and describe quality and identify who is responsible for it, they give many different responses. A popular response is "I'm not sure how to describe it, but I'll know it when I see it." Or someone might say simply that quality is a matter of meeting requirements. Perhaps the most frequent reference to quality is a remark on its absence: "How could they release something like this with such low quality!?"

These commonplace responses are telling, but they offer little room for a rigorous examination of quality and ways to improve on its execution. These comments illustrate the need to define quality so that it can be measured and achieved.

Definition

The Rational Unified Process defines quality as the characteristic of having demonstrated the achievement of producing a product that meets or exceeds agreed-on requirements—as measured by agreed-on measures and criteria—and that is produced by an agreed-on process.

Achieving quality, therefore, is not simply a matter of meeting requirements or producing a product that meets user needs or expectations. Rather, quality also includes the identification of the measures and criteria (to demonstrate the achievement of quality) and the implementation of a process to ensure that the resulting product has achieved the desired degree of quality (and can be repeated and managed).

For software development, quality should be measured for both the end product and the engineering process.

- *Product quality* is the quality of the principal product being produced (the software or system) and all the elements composing it (components, subsystems, architecture, and so on).
- *Process quality* refers to the degree to which an acceptable process (including measurements and criteria for quality) was implemented and adhered to in producing the product. Additionally, process quality is concerned with the quality of the artifacts (including iteration plans, test plans, use-case realizations, design model, and so on) produced in support of the principal product. The same concepts of product quality apply to these artifacts.

Ownership

Quality is everyone's responsibility. For this reason, there is no Worker: Quality Engineer in the Rational Unified Process. Everyone shares in the responsibility and glory for the achievement of a high-quality product (or the shame of a low-quality product). But only those directly involved in specific process workflows are responsible for the glory (or shame) for the quality of those workflows (and their artifacts).

A TAXONOMY OF TESTING

Testing is the means by which we assess quality. Testing should not be viewed as a singular activity, nor as a few individual tests. The best practice is to view testing as a broad workflow that encompasses a series of tests that are implemented throughout the development life cycle. These tests are focused on identifying and eliminating defects and assessing product quality early and continuously.

To achieve this goal, many tests, each one having a different objective or focus, should be included in the overall test workflow.

Tests can be grouped or categorized by the following:

- *Quality dimension:* the major quality characteristic or attribute that is the focus of the test
- *Stage of test:* the time during the life cycle that the test is executed
- *Type of test:* the specific test objective for an individual test, usually limited to a single quality dimension

Quality Dimension

You should address the following quality dimensions when testing software:

- *Reliability*
  Software code integrity and structure (technical compliance to language, syntax, and resource usage) and robustness (resistance to failures, such as crashes, memory leaks, and so on)
- *Function*
  Ability to execute the specified use cases as intended and required
- *Application performance*
  The timing profiles of the application, including the code's execution flow, data access, function calls, and system calls

- *System performance*
  The operational characteristics related to production load, such as response time, operational reliability (mean time to failure, or MTTF), and operational limits such as load capacity or stress

For each of the quality dimensions, one or more types of tests should be executed during one or more of the test stages.

Stages of Testing

Testing, in best practice, is not executed all at once. Testing is executed against different types of targets (targets-of-test) in different stages of the software's development. These test stages progress from testing small elements of the system, such as components (unit testing), to testing completed systems (system testing).

- *Unit test stage*
  The smallest testable elements of the system are individually tested.
- *Integration test stage*
  The integrated units (or components or subsystems) are tested.
- *System test stage*
  The complete application and system (one or more applications) are tested.
- *Acceptance test stage*
  The complete application (or system) is tested by end users (or representatives) for the purpose of determining readiness for deployment.

Types of Tests

There are many different types of tests. Each type has a specific test objective—that is, it is focused on testing only one characteristic or attribute of the software. The following are common test types.

- *Benchmark test*
  This test compares the performance of a (new or unknown) target-of-test to a known reference—workload and system.
- *Configuration test*
  This test verifies that the target-of-test functions as intended on different hardware and/or software configurations.

- *Function test*
  This test verifies that the target-of-test functions as intended, providing the required service, method, or use case to the appropriate actor.
- *Installation test*
  This test verifies that the target-of-test installs as intended on different hardware and software configurations, in the varying, required installation conditions, such as a full install, custom install, insufficient space, and so on.
- *Integrity test*
  This test verifies the reliability of the target-of-test, such as the code's technical compliance to language, syntax, and resource usage, as well as its robustness.
- *Load test*
  This test verifies and assesses the acceptability of the target-of test's operational limits under varying workloads while the system-under-test remains constant.
- *Performance test*
  This test verifies and assesses the target-of-test's performance, such as response time under a constant workload (and varying system elements) and operational characteristics (code's execution flow, data flow, and number of function calls).
- *Stress test*
  This test ensures that the target-of-test functions as intended when abnormal conditions are encountered, such as extreme workloads, insufficient memory, unavailable services or hardware, or diminished shared resources.

Regression Testing

A discussion of testing is not complete without the mention of regression testing. Regression testing is a test strategy in which previously executed tests are reexecuted against a new version of the target-of-test. The purpose of the regression test is to ensure that

- The defects identified in the earlier execution of test have been addressed
- The changes made to the code have not introduced new defects or reintroduced old ones

Regression testing can be the reexecution of any of the different test types. Typically, because testing is performed in an iterative process, regression testing is executed as part of each test effort in each iteration for each test type.

TEST MODEL

The *test model* is a representation of what in the target–of-test will be tested. It is the collection of the test cases, test procedures, test scripts, and expected test results along with a description of their relationship.

The test model consists of the following.

- *Test cases*
  The set of test data, execution conditions, and expected test results developed for a specific test objective. Test cases can be derived from use cases, design documents, or the software code.
- *Test procedures*
  The set of detailed instructions for the setup, execution, and evaluation of test results for test cases.
- *Test scripts*
  The computer-readable instructions that automate the execution of test procedures.

The relationships of the preceding artifacts are as follows.

- A test case can be implemented by one or more test procedures.
- A test procedure can implement (the whole or parts of) one or more test cases.
- A test script automates the execution of (the whole or parts of) one or more test procedures.

Figure 12-1 shows all constituents of a test model and their relationships.

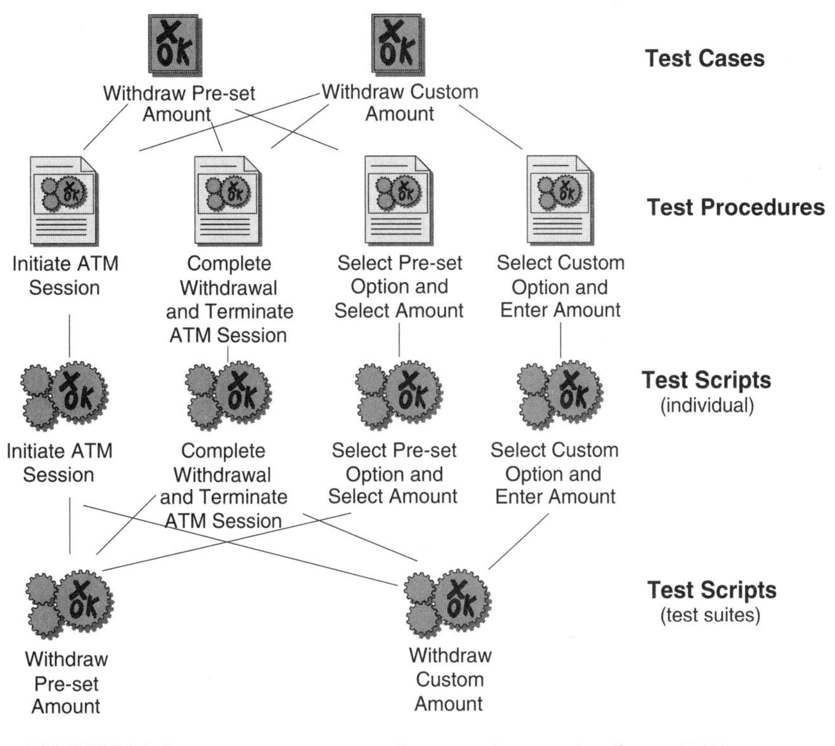

FIGURE 12-1 *Test cases, test procedures, and test scripts for an ATM system*

WORKERS AND ARTIFACTS

In the Rational Unified Process, how is all this translated concretely in terms of workers, artifacts, activities, and workflows? Figure 12-2 shows the workers and artifacts in the test workflow.

The main workers involved in the test workflow are as follows.

- The *test designer,* who is responsible for the planning, design, implementation, and evaluation of testing. This involves generating the test plan and test model, implementing the test procedures, and evaluating test coverage, results, and effectiveness.
- The *system tester,* who is responsible for executing the system tests. This effort includes setting up and executing tests, evaluating test execution, recovering from errors, assessing the results of testing, and logging identified defects.

Similarly, the following are the secondary workers.

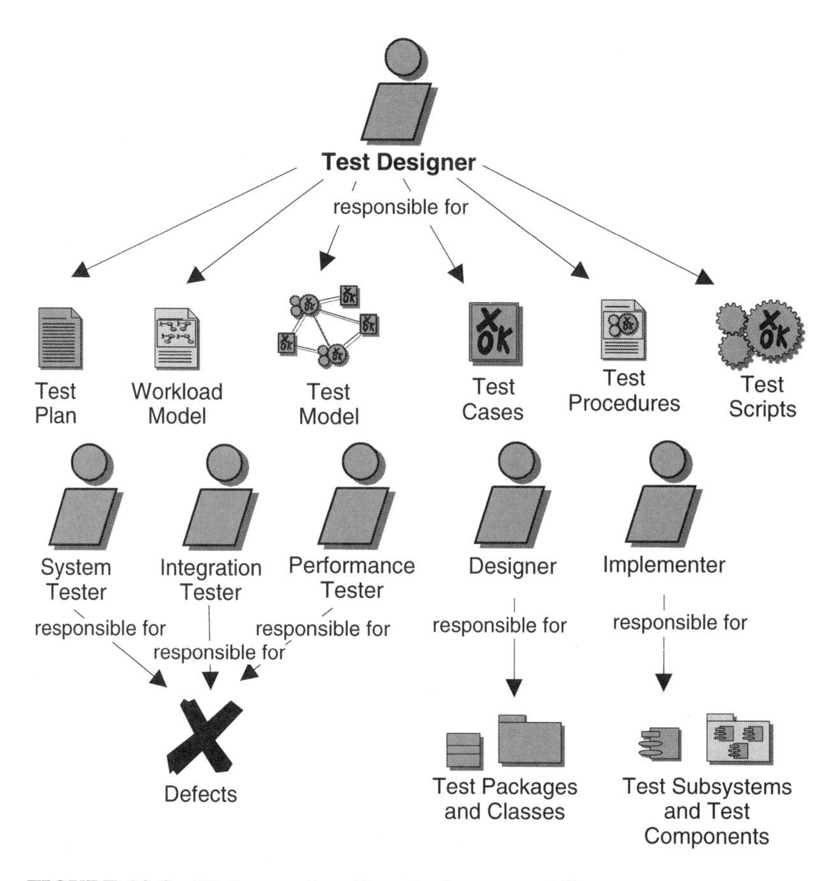

FIGURE 12-2 *Workers and artifacts in the test workflow*

- The *performance tester,* who is responsible for executing the performance tests
- The *integration tester,* who is responsible for executing the integration tests

When specific code (such as drivers or stubs) must be developed to support testing, the designer and the implementer are also involved in roles similar to the ones defined in Chapters 10 and 11.

The key artifacts of testing are as follows.

- The *test plan,* which contains information about the purpose and goals of testing within the project. The test plan identifies the strategies to be used and the resources needed to implement and execute testing.

- The *test model,* described earlier, which consists primarily of test cases, test procedures, and test scripts.
- A *workload model* for performance testing, which identifies the variables and defines their values used in the different performance tests to simulate or emulate the actor characteristics and the end user's business functions (use cases), their load, and their volume.
- *Defects* generated as a result of failed tests are one kind of change request (see Chapter 13).

In addition to these artifacts, the following artifacts are produced when test-specific code must be developed:

- Test packages and classes
- Test subsystems and components

A test evaluation report summarizing the result of testing is used as part of the project iteration assessment and periodic status assessment (see Chapter 7, The Project Management Workflow).

WORKFLOW

The test workflow is similar for each test type (see Figure 12-3) and contains these activities:

1. A planning activity
2. A design activity
3. An implementation activity
4. An execution activity
5. An evaluation activity

Planning

The test designer identifies the specific requirements for test, resource needs, and test strategies and documents them in the test plan. The test plan communicates the intent and scope of testing.

Design

In the design activity, the test designer analyzes the target-of-test and develops the test model and, in the case of performance test, the workload model.

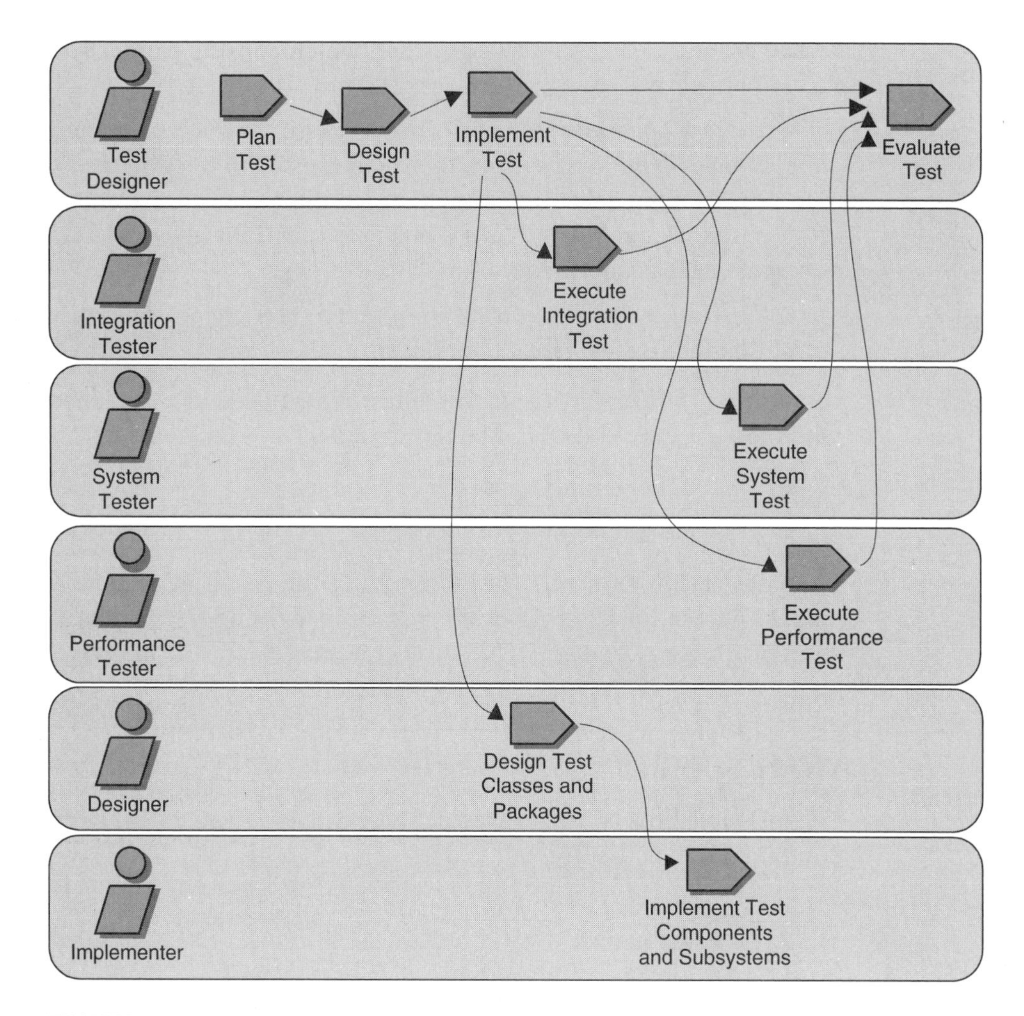

FIGURE 12-3 *A workflow in test*

Implementation

The test designer records or programs the test procedures (usually within the context of a test automation tool) that were defined in the test model. In this case, the output artifact is a test script.

If special test code (such as a test harness, drivers, or stubs) is needed to support or perform testing, the designer and the implementer work with the test designer to design and implement the test code.

Execution

The tester executes the tests against the target-of-test (manually by using the test procedure or automatically by executing the test script). When the tests are completed, the tester reviews the results to determine whether testing was executed properly and then documents any discovered defects. These defects are then submitted for review and analysis.

Evaluation

The test designer assesses the results of test and delivers quantifiable measures of testing used to determine the quality of the target-of-test and the efficiency of the test process.

TOOL SUPPORT

As stated earlier, testing is an iterative effort that is implemented throughout the development life cycle. To maximize the efficiencies and effectiveness of testing, test automation and test tools are a necessity. This is especially true for some types of tests (such as performance or load tests) in which thousands of machines and testers must be synchronized.

The Rational Suite of software development tools supports test automation and the test workflow as follows.

- TestStudio implements and executes GUI-based test scripts to test reliability, function, and application performance. TestStudio includes the following tools.
  - Requisite Pro manages and tracks requirements for test.
  - Purify identifies runtime errors for reliability tests.
  - Quantify/Visual Quantify identifies and reports on the operational characteristics of the software, thereby enabling developers and testers to identify and eliminate performance bottlenecks.
  - PureCoverage/Visual PureCoverage identifies and reports code coverage (during test execution).
  - ClearQuest manages defects and produces defect reports (to evaluate product and process quality).

- PerformanceStudio implements and executes virtual user (VU) test scripts for performance and limited functional tests.

The Rational Unified Process provides tool mentors for many of these tools.

SUMMARY

- Product quality is the quality of the principal product being produced (the software or system) as well as that of all the elements that compose the product.
- Process quality is the implementation of a process and the measurement of adherence to the process, its standards, and guidelines.
- Quality is everyone's responsibility.
- The test workflow is aimed primarily at assessing product quality in terms of interactions between components, integration, functionality, and performance.

Chapter 13

The Configuration and Change Management Workflow

with Jas Madhur

THIS CHAPTER INTRODUCES KEY concepts in software configuration management and then describes the workers, activities, and artifacts involved in the configuration and change management workflow.

PURPOSE

The purpose of the configuration and change management workflow is to track and maintain the integrity of project assets as they evolve in the presence of changes.

During the course of the development life cycle, many valuable artifacts are created. Developing these artifacts is labor-intensive, and they represent a significant investment. As such they are an important asset that must be safeguarded and easily available for reuse. These artifacts evolve and, especially in an iterative development, are updated again and again. Although one worker usually is responsible for the artifact, we cannot rely on the individual's memory (or upper-left desk drawer) to keep track of these project assets. The project team members must be able to identify and locate artifacts, to select the appropriate version of an artifact, to look at its history to understand its current state and the reasons it has changed, and to ascertain who is currently responsible for it.

At the same time, the project team must track the evolution of the product, capture and manage requests for changes no matter where they come from, and then implement the changes in a consistent fashion across sets of artifacts.

Finally, to support the project management workflow, we must provide status information on the key project artifacts and gather metrics related to their changes.

THE CCM CUBE

Configuration and change management (CCM) covers three interdependent functions. These key aspects can be illustrated using the "CCM cube" shown in Figure 13-1. The cube has three faces, and each face looks at a different aspect of the problem. All three aspects are deeply intertwined.

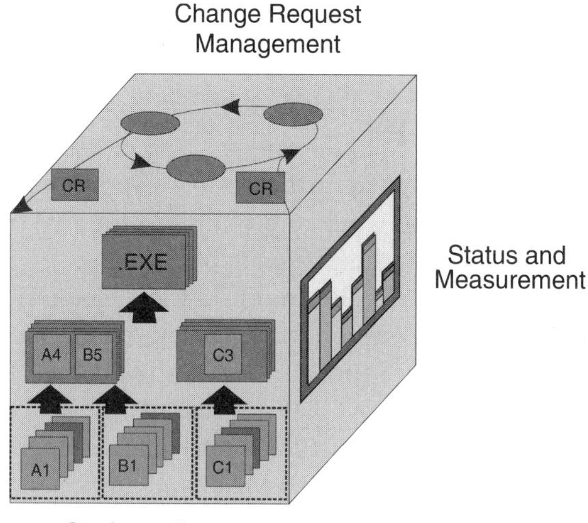

FIGURE 13-1 *The CCM cube*

- The *configuration management face is related to the product structure.*
  Configuration management (CM) deals with the issue of artifact identification, versions, and dependencies between artifacts as well as the identification of configurations that are consistent sets of interrelated artifacts. It also deals with the issue of providing workspaces to individuals and teams so that they can develop without constantly stepping on one another's feet.
- The *change request management face is related to the process structure.*
  Change request management (CRM) deals with the capture and management of requested changes generated by internal and external stakeholders. It also deals with the analysis of the potential impact of the change and with the tracking of what happens with the change until it is completed.
- The *status and measurement face is related to the project control structure.*
  Status and measurement deals with the extraction of information for project management from the tools that support the configuration management and the change request management functions. The following information is useful for an assessment:
  - The product's status, progress, trends, and quality
  - What has been done and what remains to be done
  - The expenditures
  - The problem areas that require attention

Now let's examine each of these three aspects in more detail.

Configuration Management

The configuration management aspect of this workflow deals with the product structure, as shown in Figure 13-2. Important artifacts are placed under version control. As an artifact evolves, multiple versions exist, and you must identify the artifact, its versions, and its change history.

Artifacts depend on one another. As a result, each change causes a ripple effect. When an artifact has been modified, dependent artifacts may have to be revisited, modified, or regenerated.

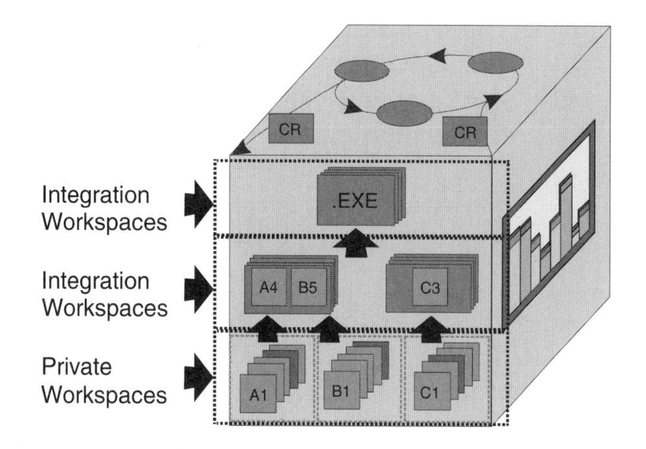

FIGURE 13-2 *Product structure and workspaces*

For example, a C++ source file depends on the header files it includes. It also depends on the class specification it implements.

The latest version of an artifact often is the best one, but there are many circumstances in which things are not that simple: multiple developers may work in parallel on the same artifact, or variants of the same artifacts may be required for different end products.

The knowledge of the dependencies between artifacts, the transformations that occur along the dependencies, and perhaps the tool used to effect these transformations can be exploited to recreate stacks of dependent artifacts. For example, a makefile can be created to trigger the recompilation and linkage of an entire application based on a few modified source files.

Build management is about making sure that components that would constitute a build are available and are assembled in the right order based on their dependency hierarchies. Effective build management requires that we can trust that constituent components have been adequately developed and tested to be included in a build. This is akin to the notion of a pedigree: An artifact is OK only if its ancestors are OK.

A workspace provides a "sandbox" in which individual developers or small teams work as if they were isolated from one another. It provides access to a necessary set of artifacts. Workspaces can be used for basic development or for the progressive integration of products, playing the role of staging areas.

The overall product structure—the organization of all artifacts—is driven by the implementation view of the architecture.

Change Request Management

Change request management deals with the process structure and is shown at the top of the cube in Figure 13-3. A change request is a documented proposal of a change to one or more artifacts. Change requests can be raised for a variety of reasons: to fix a defect, to enhance product quality (such as usability or performance), to add a requirement, or simply to document the delta between one iteration and the next.

Change requests have a life represented as a simple state machine, with states such as new, logged, approved, assigned, and complete. As the change requests go from state to state, information about the change is added, such as the reason for the change, the motivation, the affected artifacts (artifacts that must be modified simultaneously), and the impact on the design, the architecture, the cost, and the schedule. Not all change requests are acted on. A management decision must take into account the result of the impact analysis as well as relative priorities to determine which changes will be implemented, when they will be implemented, and in which release they will be implemented.

To perform a given change, the exact workflow is driven by the type of artifacts affected and their dependencies. If a change affects

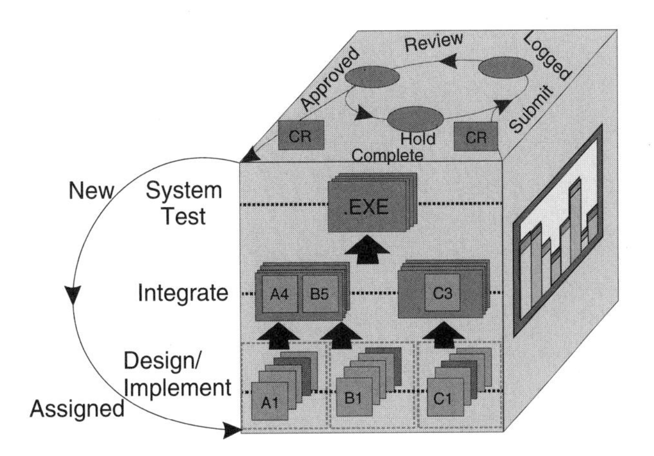

FIGURE 13-3 ***Change requests***

the design of a class, the workflow will include the Activity: Design Class and the Worker: Designer. The workflow may also include any activity and worker affected downstream, such as Activity: Review Class, Activity: Database Design, Activity: Implement Class, and so on. In a sense, change requests drive the actual process.

Changes to artifacts are tracked and associated with the corresponding change request. Change requests are closed when the change has been completed, tested, and included in a release.

Status and Measurement

Status and measurement deals with the project control structure and is shown on the right face of the cube in Figure 13-4. Change requests have a state as well as other attributes such as root cause, nature (such as defect or enhancement), severity, priority, and area affected (such as layer or subsystem). The various states of a change request provide useful tags as the basis for metrics reporting.

Many change requests are accumulated in a change request database, and they represent the bulk of the work to be done during an iteration, especially in the construction and transition phases.

Status information on the progress of the project can be extracted from this database:

- Overall progress of the project relative to the changes
- Number of changes made in the various states

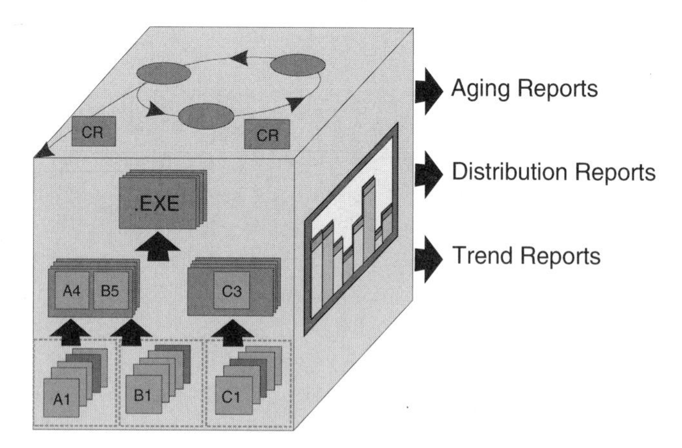

FIGURE 13-4 ***Status and measurement***

- Age of the change requests—that is, how long they have been in a particular state

We can also extract reports on the *distribution* according to the following:

- By severity or priority
- By layer or subsystem affected
- By team
- By root cause

Trend reports look at the derivative of these metrics over time, a measure that is useful for projection.

WORKERS AND ARTIFACTS

In the Rational Unified Process, how is all this translated concretely in terms of workers, artifacts, activities, and workflow? Figure 13-5 shows workers and artifacts in the configuration and change management workflow. The main workers involved in configuration and change management are described on the next page.

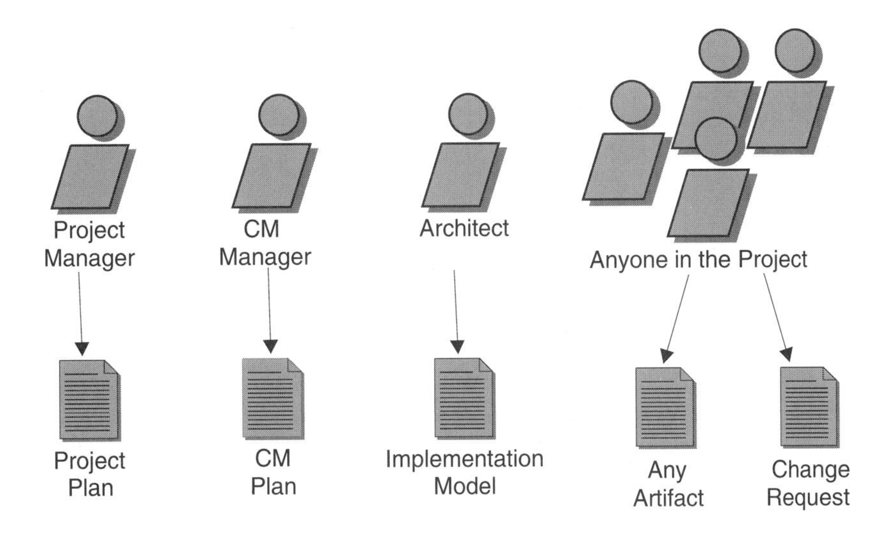

FIGURE 13-5 ***Workers and artifacts***

- The *project manager is* responsible for the configuration management plan, one of the components of the overall software development plan (SDP). The project manager is also the recipient and user of the status and measurement reports.
- The *configuration manager* is responsible for setting up the product structure in the CM system; for defining and allocating workspaces for developers; and for integration. The configuration manager also extracts the appropriate status and metrics reports for the project manager.

The following workers are also involved in this workflow.

- *Implementers* access adequate workspace and the artifacts they need to implement the changes for which they are responsible.
- *Integrators* accept changes in the integration workspace and build the product.
- *Any worker* can submit a change request.
- The *architect* provides input to the product structure by means of the implementation view.

The *Change Control Board* (CCB) is a group composed of various technical and managerial stakeholders, such as the project manager, the architect, the configuration manager, and any stakeholder (customer representative, marketing personnel, and so on). The role of the CCB is to assess the impact of changes, determine priorities, and approve changes.

The key artifacts of configuration and change management are as follows.

- *The configuration management plan*
  The CM plan describes the policies and practices to be used on the project for CM: versions, variants, workspaces, and procedures for change management, builds, and releases. The CM plan defines the rules and responsibilities for the CCB. It is a part of the software development plan.
- *Change requests*
  Change requests may be of a wide variety: they may document defects, changes to requirements, or the delta from one iteration to the next. Each change request is associated with an originator and a root cause. Later, impact analysis

attaches the impact of the change in terms of affected artifacts, cost, and schedule. As the change request evolves, its state changes; history items, in particular the CCB decisions, are attached to it.

In addition, this workflow includes the following.

- The *implementation model* drives the product used by the configuration manager to set up the CM environment.
- *Metrics and status reports* are extracted from the CM and CRM support environment to be included in status assessment reports.

WORKFLOW

There are two interwoven workflows in configuration and change management: one from the perspective of the CM product structure, shown in Figure 13-6, and a second one from the perspective of the life of a change request.

1. The project manager establishes the concrete CM practices to be used in this project and captures them in the CM plan.
2. Based on this plan and using the input of a software architecture document (the implementation view), the configuration manager establishes the product structure and creates and allocates the various workspaces that are required by the developers and integrators.
3. Workers access their workspace and modify artifacts; in particular, they implement changes allocated to them.
4. Integrators accept changes in integration workspaces and build partial products to be tested.
5. A new baseline is cut—for example, at the end of an iteration.

From the point of view of a change request, the workflow is slightly different.

1. A change request is entered.
2. It is analyzed for its impact on artifacts, cost, and schedule.
3. The CCB reviews the impact analysis, prioritizes changes, and approves them.
4. The changes are allocated to various workers for implementation.

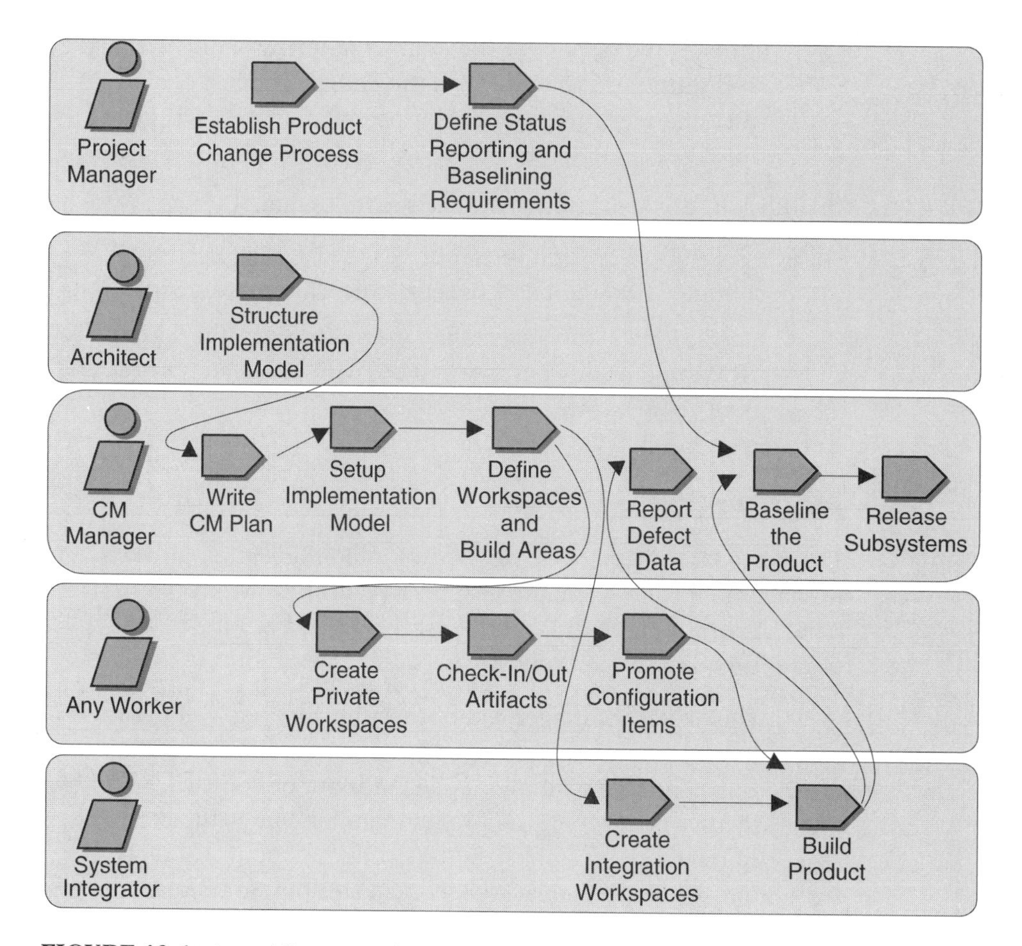

FIGURE 13-6 *A workflow in configuration management*

5. Changes are incorporated in a build and tested.
6. The changes are closed.

TOOL SUPPORT

Managing all this change is hard. The complexity of doing it right, day after day, grows exponentially with the size of the team, the size of the product (that is, the number of artifacts), the geographic distribution of the team, and the schedule pressure. It is error-prone, tedious, and labor-intensive and is therefore an obvious task that can benefit from tool support and automation.

The Rational Suite of software development tools supports the configuration and change management workflow with the following tools.

- ClearCase assists with the configuration management portion.
- ClearQuest assists with the change request management and status and measurement portion.

The Rational Unified Process provides tool mentors for both tools.

SUMMARY

- The purpose of the configuration and change management workflow is to maintain the integrity of the project artifacts as they evolve in the presence of change requests.
- Configuration management deals with the product structure, identification of elements, version, valid configuration of elements, and workspaces.
- Change request management encompasses the process of modifying artifacts in a consistent fashion.
- Status and metrics can be extracted from configuration and change management information to assist in status assessment.
- Tools such as ClearCase and ClearQuest automate the most tedious aspects of this workflow.

Chapter 14
The Environment Workflow

THIS CHAPTER INTRODUCES the kinds of activities that take place in the environment workflow.

PURPOSE

The purpose of the environment workflow is to support the development organization with both processes and tools. This support includes the following:

- Tool selection and acquisition
- Toolsmithing—that is, adjusting the tools to suit the organization or building additional tools
- Process configuration
- Process improvement
- Training
- Technical services to support the process: the information technology (IT) infrastructure, account administration, backup, and so on

Some of the activities related to process implementation and configuration are described in Chapter 17.

WORKERS AND ARTIFACTS

Figure 14-1 summarizes the workers and artifacts involved in the environment workflow. The main worker involved with the process is the *process engineer,* who is responsible for the software development process itself. This includes configuring the process before project start-up and continuously improving the process during the development effort. The main artifact is the *development case,* which specifies the tailored process for the individual project.

In certain aspects of the process, the process engineer needs the competence of other workers to establish the *guidelines*:

- A *business process analyst,* for the business modeling guidelines
- A *system analyst,* for the use-case modeling guidelines
- A *user-interface designer,* for the user-interface guidelines
- An *architect,* for the design guidelines
- A *technical writer,* for the user manual style guide

The following main workers are involved with the tool environment.

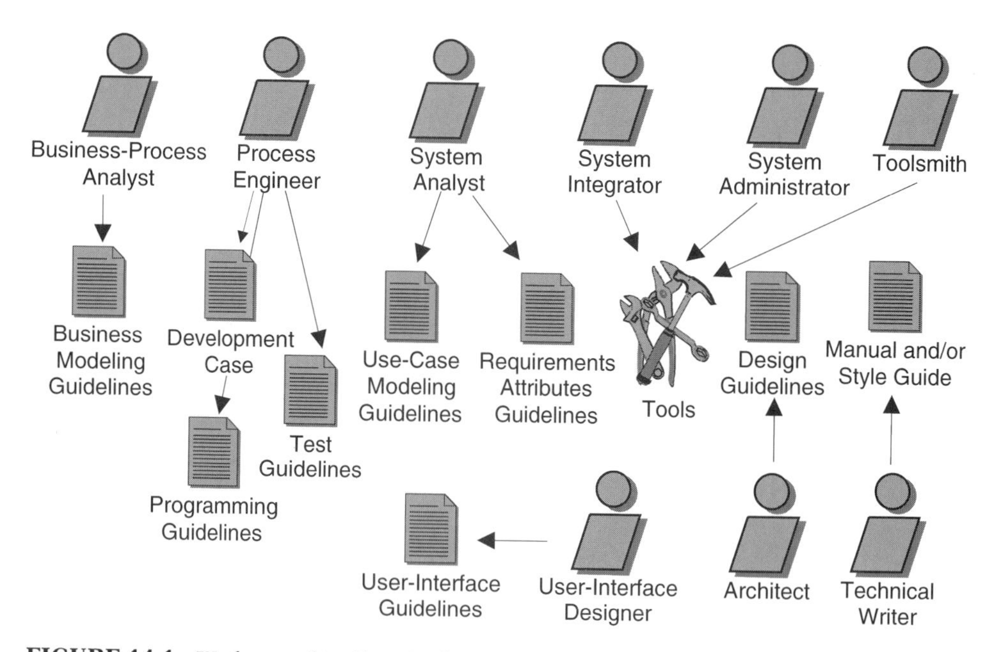

FIGURE 14-1 *Workers and artifacts in the environment workflow*

- The *toolsmith* develops tools to support special needs, to provide additional automation of tedious or error-prone tasks, and to provide better integration between tools.
- The *system administrator* maintains the hardware and software development environment and performs system administrative tasks such as account administration, back-ups, and so on.

The main artifact is the *software development environment:* hardware, software, network resources, software tools, and support software for development and test activities.

WORKFLOW

Now let's look at a few typical activities that may take place in the environment workflow.

Configuring the Process

This activity consists of adapting the process product to the needs and constraints of the adopting organization by modifying the process framework delivered by Rational software. See Chapter 17 for more details.

Implementing the Process

Implementing the Rational Unified Process in a software-development organization means changing the organization's practice so that it routinely and successfully uses the Rational Unified Process in part or in whole. See Chapter 17 for more details.

Selecting and Acquiring Tools

Many of the steps in the process can be efficiently performed only with the proper tool support. Tools must be selected that fit the particular needs of an organization, based primarily on specific activities or artifacts necessary for the process. The following wide palette of tools may be needed:

- Tools for modeling
- Tools for requirements management
- Tools for code development (editors, compilers, debuggers)
- Tools for configuration management and change management
- Tools for testing
- Tools for planning and tracking
- Tools for documentation preparation

The tools must be acquired, installed, and configured. The user community must be trained to use them. Specific project guidelines specify their usage: file naming conventions, procedures, scripts to automate certain tasks, and so on. As discussed in Chapter 3, tool mentors can be added to the process to explain how the tool must be used to fulfill certain activities or steps.

Toolsmithing

Sometimes, special tools must be developed internally to support special needs, provide additional automation of tedious or error-prone tasks, and provide better integration between tools. The Rational Unified Process can also be used for this activity of tool development, although with a lighter-weight process than the one used for developing the product.

Supporting the Development

This activity encompasses a broad range of technical services, such as maintaining the hardware and software development environment; performing system administration, backups, and upgrades; keeping the telecommunication system operational; and creating and reproducing documents.

Training

In some cases, a development organization may be large enough to justify having its own internal training organization to educate the developers on the process, tools, and techniques to be used on the project.

SUMMARY

- The goal of the environment workflow is to provide adequate support for the development organization in tools, processes, and methods.
- Many activities and steps of the Rational Unified Process can be automated through the use of tools, thereby removing the most tedious, human-intensive, and error-prone aspects of software development.

Chapter 15
The Deployment Workflow

THIS CHAPTER GIVES AN overview of the kinds of activities that take place in the deployment workflow.

PURPOSE

The purpose of the deployment workflow is to deliver the product to the end users. This workflow involves a wide range of activities, such as the following:

- Producing and assembling a complete external release of the software
- Packaging the software
- Distributing the software
- Installing the software
- Training the users or the sales force
- Providing help and assistance to users
- Planning and conducting beta tests
- Migration of existing software or data
- Formal acceptance

Deployment is highly dependent on your specific domain and business context, and you must configure and tailor the Rational Unified Process to capture the activities and steps of your specific deployment workflow.

WORKERS AND ARTIFACTS

The following are examples of workers who may be involved in the deployment workflow.

- The *deployment manager* plans and organizes the deployment.
- The *implementer* produces the software part of the release.
- The *technical writer* produces the user manual, the installation notes, and so on.
- The *course developer* produces any training materials.

The artifacts may include the following:

- A *deployment plan*
- *User manuals*
- *Training materials*—slides, online tutorial, or computer-based training (CBT)

Figure 15-1 shows the workers and artifacts in the deployment workflow.

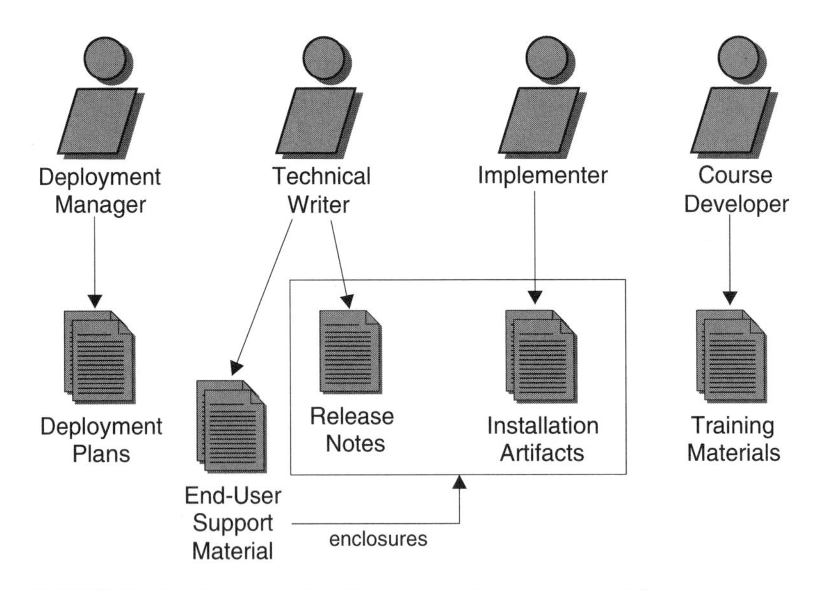

FIGURE 15-1 *Workers and artifacts in the deployment workflow*

WORKFLOW

Let's look at a few typical activities that may take place in the deployment workflow.

Producing the Software

The output of the implementation and test workflows is tested executables. To constitute a complete product, these executable programs must be associated with other artifacts:

- Installation scripts
- User documentation
- Configuration data
- Additional programs for migration: data conversion

In some circumstances, different executables need to be produced for different user configurations. Or different sets of artifacts must be assembled for different categories of users (new users versus existing users, variants by country or by language, and so on).

For distributed software, different sets may have to be produced for different computing nodes in the network.

Packaging the Software

The various artifacts that constitute the delivered product are packaged on suitable media—diskettes, tapes, CD-ROM, archived server files, books, videotapes, and so on—and should be properly identified and labeled. The activities often involve dealing with external organizations to package the software.

Distributing the Software

Again, there is a wide range of options, from packing in shipping boxes to using a network of distributors to Internet distribution (often referred to as "Web commerce").

One issue is licensing: controlling who is authorized to use the software. Software licensing usually involves setting up procedures and tools to manage licenses and communicate license codes to the users.

Installing the Software

With the advent of Internet distribution, software installation is becoming an increasingly user-controlled process. It must, however, be supported by installation tools and procedures delivered with the product. In rarer cases (large, complex technical systems), installation is performed by the software vendor.

Installation is generally more complex in the case of a distributed system, in which all nodes must be brought up-to-date in a timely fashion and the installation may be divided into multiple installation procedures.

Migration

Migration often is an issue in installation and may include the following:

- Replacing an older system with a new one, with or without constraints regarding continuity of operation
- Converting existing data to a new format

The programs associated with this migration (converters) are developed and tested using exactly the same process that is used for the primary product.

Providing Assistance to the Users

User assistance can take various forms:

- Formal training courses
- Computer-based training
- Online guidance and help
- Telephone support
- Internet support
- Collateral materials—tips, application notes, examples, wizards, and so on

Support often involves setting up procedures for problem tracking and resolution, something that integrates with the change management activities (see Chapter 13).

Acceptance

In some contractual development circumstances, the deployment includes formal acceptance by the customer of the delivered software.

Planning and Conducting Beta Tests

Many products are tested as part of deployment in the early stage of the transition phase by a representative community of users. A subset of the product is delivered to this user community, and procedures are put in place to solicit, capture, and analyze their feedback.

SUMMARY

- The deployment workflow takes care of all artifacts that end up being delivered to the end users or customers as well as to the supporting organizations: support, marketing, distribution, and sales.
- This workflow is highly dependent both on the kind of product being developed and on the business context, and it must be specialized by the organization that is adopting the Rational Unified Process.

Chapter 16
Iteration Workflows

THIS CHAPTER OFFERS a high-level description of three typical iteration workflows, one for each of the first three phases of the process.

PURPOSE

The core engineering workflows presented in sequence from Chapter 8 to Chapter 13 may wrongly give an impression of a waterfall process. Remember that they are archetypal workflows that are designed to give an overview of all activities.

The concrete workflows of a project will vary depending on the nature of the project and where you are in the life cycle. Remember, too, that these workflows are revisited again and again at each iteration.

In Chapter 3 we introduced the concept of iteration workflows, which are the more concrete workflows a project goes through during one iteration. In this chapter, we give three examples of iteration workflows:

- An iteration in the inception phase to define the project vision and the business case
- An iteration early in the elaboration phase to build an architectural prototype
- An iteration late in the construction phase to implement the system

These examples are sketchy and do not involve all activities from all workflows.

DEFINING THE PRODUCT VISION AND THE BUSINESS CASE

The context of this first example is the initial development cycle for a new product (see Figure 16-1). It illustrates a "greenfield" development for a small system.

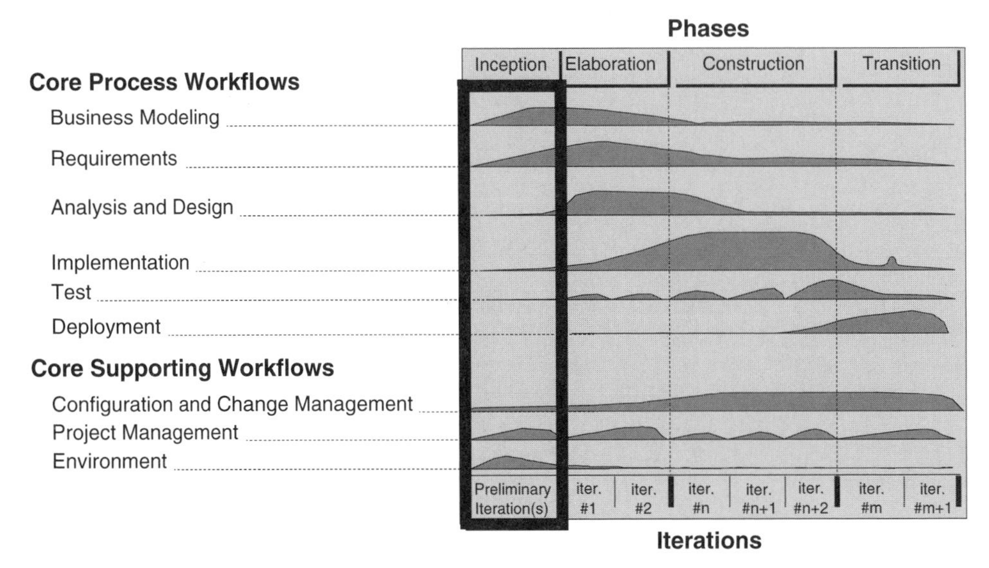

FIGURE 16-1 *An iteration in the inception phase*

Start-up: Define the System's Vision and Scope

The stakeholders of the system to be developed (such as customers, managers, marketing, and funding authorities) work with the system analysts to define the vision and scope of the project (Artifact: Vision Document). A primary aspect to consider is the users' needs and expectations (Artifact: Stakeholders' Needs). Also considered are constraints on the development project, such as platforms to be supported and external interfaces. Based on the early sketches of the vision, the group starts to prepare the business case and initiate a risk list, enumerating any hazards that could arise on the way to success (Artifacts: Business Case and Risk List).

Requirements: Outline and Clarify the System's Functionality

The system analysts use various techniques, such as storyboarding and brainstorming, to conduct sessions to collect stakeholders' opinions about what the system should do. Together, they sketch an initial outline of the system's use-case model (Artifact: Use-Case Model). An initial glossary is also captured to simplify the maintenance of the use-case model and to keep it consistent (Artifact: Glossary). The main results of these sessions are a stakeholder needs document and possibly an outline of a use-case model: a use-case survey.

Project Management: Consider the Project's Feasibility and Outline the Project Plan

With input from the use-case modeling, it is time to transform the vision into economic terms by considering the project's investment costs, the resource estimates, the development environment, the success criteria (revenue projection and market recognition), and so on, and to initiate a business case. The risk list is updated, taking into account business risks. The project manager builds a phase plan showing tentative dates for the inception, elaboration, construction, and transition phases along with the major milestones. The software development plan envisages the development environment and process that will be needed (Artifact: Project Plan).

Project Management: Refine the Project Plan

At this stage, the stakeholders should have a good understanding of the product's vision and feasibility. An order of priority among features and use cases is established. The project manager may start planning the project in more detail, basing the project plan on the prioritized use cases and associated risks. A tentative set of iterations is defined, with objectives for each iteration. The results achieved at this stage are refined in each subsequent phase and iteration and become increasingly accurate as iterations are completed.

Result

The result of this initial iteration is a first cut at the project's vision, business case, and scope, as well as the project plan. The stakeholders (organization) initiating the project should have a good understanding of the project's return on investment (ROI)—that is, what is returned at what investment costs? Given this knowledge, a "go/no go" decision can be made.

Project schedules are not written in stone. This is a key differentiator in using this process: the recognition that initial project plan estimates are rough estimates that become more realistic as the project progresses and real metrics are developed on which to base estimates. With project planning, practice makes perfect. It's important to stress continual refinement of the project plan.

Subsequent Iterations in Inception

You can initiate subsequent iterations to further enhance the understanding of the scope of the project and the system to be developed.

This activity might imply a further enhancement of the use-case model or of the plans, staffing, environment, and risks. The need for such additional work depends on variables such as the complexity of the system, the associated risks, and the domain experience.

BUILDING AN ARCHITECTURAL PROTOTYPE

Figure 16-2 illustrates our second example: an iteration early in the elaboration phase. The context of this example is that the inception phase has been completed and the life-cycle objective milestone has been passed; we have an outline of the actors and use cases as well as an initial cut at a project plan. (See the iteration workflow discussed in the preceding section.)

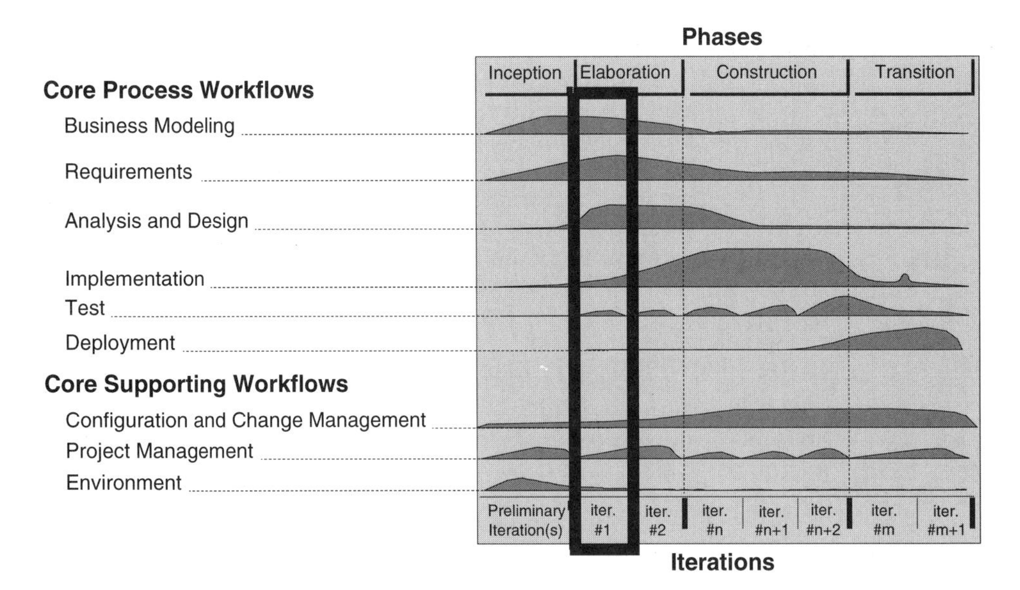

FIGURE 16-2 *An iteration early in the elaboration phase*

Start-up: Outline the Iteration Plan, Risks, and Architectural Objectives

Based on the project plan outlined earlier, the project manager starts sketching an iteration plan for the current iteration (Artifact: Iteration Plan). Evaluation criteria for the architecture are outlined in discussions with the architect and by considering the architectural risks to be mitigated (Artifact: Risk List). Remember that one of the goals of elaboration is to establish a robust, executable architecture; the plan for doing this must be developed in this initial elaboration iteration.

Requirements: Decide Which Use Cases and Scenarios Will Drive the Development of the Architecture

The architect continues by discussing an initial use-case view with the project manager to determine which use cases and scenarios should be focused on in this iteration; these use cases and scenarios will drive the development of the architecture. Note that the identification of these use cases and scenarios affects the iteration plan described in the preceding step, and the iteration plan should be updated.

Requirements: Understand This Driver in Detail and Inspect the Results

A number of use-case specifiers describe some of the selected use cases and scenarios in detail: the higher-priority, most critical, and most complex ones to be addressed in this first elaboration iteration. The system analyst might then need to restructure the use-case model as a whole. The changes to the use-case model are then reviewed and approved (Artifact: Use-Case Model and Supplementary Specification).

Reconsider Use Cases and Risks

The architect revisits the use-case view, taking into consideration new use-case descriptions and possibly a new structure of the use-case model. The task now is to select the set of use cases and scenarios to be analyzed, designed, and implemented in the current iteration. Note again that the development of these use cases and scenarios sets the software architecture. The project manager again updates the current iteration plan accordingly (Artifact: Iteration Plan) and might reconsider risk management because new risks might have been made visible according to new information (Artifact: Risk List).

Requirements: Prototype the User Interface

Using the use cases that have already been fleshed out, a user-interface designer starts expanding them in use-case storyboards and builds a user-interface prototype to get feedback from prospective users (Artifact: User-Interface Prototype).

Analysis and Design: Find Obvious Classes, Do Initial Subsystem Partitioning, and Look at Use Cases in Detail

To get a general sense of the obvious classes needed, the architect considers the system requirements, the glossary, the use-case view (but no use cases), and the team's general domain knowledge to sketch the outline of the subsystems, possibly in a layered fashion. The architect also identifies the analysis mechanisms that constitute common solutions to common problems during analysis. The software architecture document is initiated (Artifact: Software Architecture Document). In parallel with this effort, a team of designers, possibly together with the architect, starts finding classes or objects for this iteration's use cases or scenarios. This team also begins to allocate responsibilities to the identified classes and analysis mechanisms.

Analysis and Design: Refine and Homogenize Classes and Identify Architecturally Significant Ones; Inspect Results

A number of designers refine the classes identified in the preceding step by allocating responsibilities to the classes and updating their relationships and attributes. It is determined in detail how the available analysis mechanisms are used by each class. Then the architect identifies a number of classes that should be considered architecturally significant and includes them in the logical view (Artifact: Software Architecture Document).

Analysis and Design: Consider the Low-Level Package Partitioning

To analyze the service aspect of the architecture, the architect organizes some of the classes into design packages and relates these packages to the subsystems. The subsystem partitioning might need to be reconsidered.

Analysis and Design: Adjust to the Implementation Environment, Decide the Design of the Key Scenarios, and Define Formal Class Interfaces; Inspect Results

The architect then refines the architecture by deriving the design mechanism needed by the earlier identified analysis mechanisms. The design mechanisms are constrained by the implementation environment, that is, the implementation mechanisms available: operating system, middleware, database, and so on. Designers instantiate these classes into objects when they describe how the selected use cases or scenarios are realized in terms of collaborating objects in interaction diagrams. This puts requirements on the employed classes and design mechanisms; the interaction diagrams previously created are refined. Given the detailed requirements that are then put on each object, the designers merge these into consistent and formal interfaces on their classes. The requirements put on each design mechanism are handled by the architect, who updates the logical view accordingly. The resulting design artifacts are then reviewed.

Analysis and Design: Consider Concurrency and Distribution of the Architecture

The next step for the architect is to consider the concurrency and distribution required by the system. The architect studies the tasks and processes required and the physical network of processors and other devices. An important input to the architect here is the designed use cases in terms of collaborating objects in interaction diagrams: the use-case realizations (Artifact: Software Architecture Document).

Analysis and Design: Inspect the Architectural Design

The architecture is reviewed.

Implementation: Consider the Physical Packaging of the Architecture

An architect now considers the impact of the architectural design on the implementation model and defines the implementation view.

Implementation: Plan the Integration

A system integrator studies the use cases that are to be implemented in this iteration and defines the order in which subsystems should be implemented and later integrated into an architectural prototype. The results of this planning should be reflected in the project plan (Artifact: Project Plan).

Test: Plan Integration Tests and System Tests

A test designer plans the system tests and the integration tests, selecting measurable testing goals to be used when assessing the architecture. These goals could be expressed in terms of the ability to execute a use-case scenario with a certain response time or under specified load. The test designer also identifies and implements test cases and test procedures (Artifact: Test Plan).

Implementation: Implement the Classes and Integrate

A number of implementers code and unit-test the classes identified in the architectural design. The implementations of the classes are physically packaged into components and subsystems in the implementation model. The integration testers test the implementation subsystem, and then the implementers release the subsystems to integration.

Integrate the Implemented Parts

The system integrators incrementally integrate the subsystems into an executable architectural prototype. Each build is tested.

Test: Assess the Executable Architecture

Once the whole system (as defined by the goal of this iteration) has been integrated, the system tester tests the system. The test designer then analyzes the results to make sure that the testing goals have been reached. The architect then assesses this result and compares it with the risk identified initially.

Assess the Iteration Itself

The project manager compares the iteration's actual cost, schedule, and content with those of the iteration plan; determines whether rework needs to be done and, if so, assigns it to future iterations; updates the risk list (Artifact: Risk List); updates the project plan (Artifact: Project Plan); and prepares an outline of an iteration plan for the next iteration (Artifact: Iteration Plan). Other lessons learned in terms of productivity, process improvement, tool support, and training are also interesting to consider at this stage, and actions are defined for the next iteration.

Result

The result of this initial iteration is a first cut at the architecture. It consists of fairly well described architectural views (the use-case view, the logical view, and the implementation view) and an executable architectural prototype.

Subsequent Iterations in Elaboration

Subsequent iterations further enhance the understanding of the requirements and of the architecture. This might imply a further enhancement of the design or implementation model (that is, the realization of more use cases in priority order).

IMPLEMENTING THE SYSTEM

The context of our third example is that in this late construction phase, requirements are stable and a great deal of the functionality has been implemented and integrated in preparation for test. The project is approaching its first beta release and the initial operational capability milestone (see Figure 16-3).

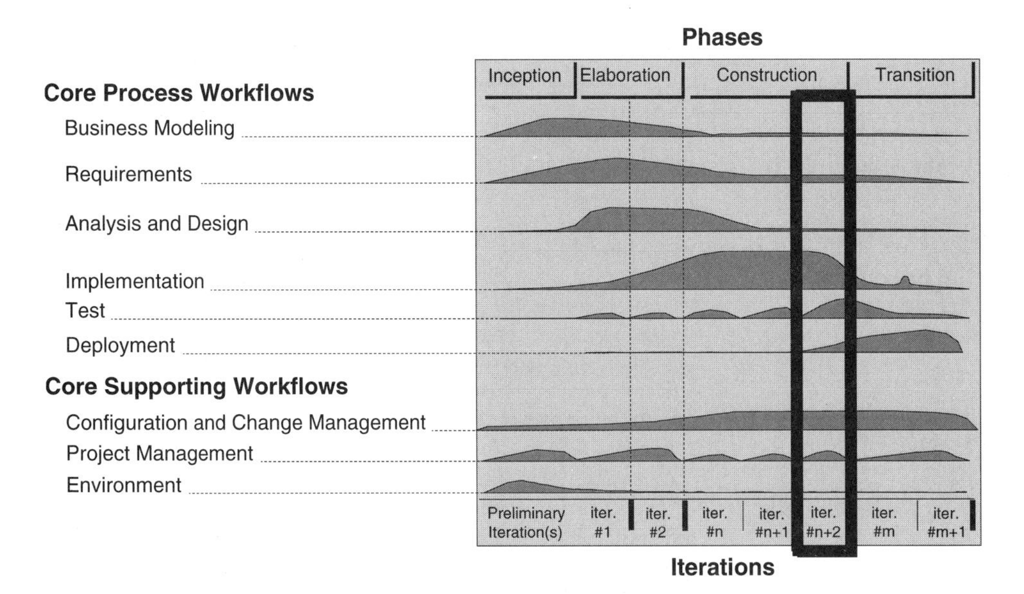

FIGURE 16-3 *An iteration late in construction*

Project Management: Plan the Iteration

The project manager updates the iteration plan based on any new functionality that is to be added during the new iteration, factoring in the current level of product maturity, lessons learned from the previous iterations, and any risks that need to be mitigated in the upcoming iteration (Artifact: Iteration Plan and Artifact: Risk List).

Implementation: Plan System-Level Integration

Integration planning takes into account the order in which subsystems are to be put together to form a working and testable configuration. The choice depends on the functionality already implemented and on the aspects of the system that must be in place to support the overall integration and test strategy. This is done by the system integrator, and the results are documented in the build plan (Artifact: Integration Build Plan). The integration build plan defines the frequency of builds and describes when given build sets will be required for ongoing development, integration, and test.

Test: Plan and Design System-Level Test

The test designer ensures that there will be an adequate number of test cases and procedures to verify testable requirements. The test designer must identify and describe test cases and identify and structure test procedures in the test model. In general, each test case has at least one associated test procedure. The test designer also reviews the accumulated body of tests from preceding iterations, which could be modified to be reused in regression testing for current and future iteration builds (Artifact: Test Scripts and Test Procedures).

Analysis and Design: Refine Use-Case Realizations

Designers refine the classes identified in previous iterations by allocating responsibilities to specific classes and updating their relationships and attributes. New classes may also need to be added to support possible design and implementation constraints. Changes to classes may require a change in subsystem partitioning.

Test: Plan and Design Integration Tests at the Subsystem and System Levels

Integration tests focus on the degree to which developed components interface and function together. The test designer follows the test plan that describes the overall test strategy, required resources, schedule, and completion and success criteria. The designer identifies the functionality that will be tested together and the stubs and drivers that must be developed to support the integration tests. The implementer develops the stubs and drivers based on the input from the test designer.

Implementation: Develop Code and Test Unit

In accordance with the project's programming guidelines, implementers develop code to implement the classes in the implementation model. They fix defects and provide any feedback that may lead to design changes based on discoveries made in implementation.

Implementation: Plan and Implement Unit Tests

The implementer designs unit tests that address what the unit does (the black-box test) and how it does it (the white-box test). Under black-box (specification) testing, the implementer must be sure that the unit, in its various states, performs to its specification and can correctly accept and produce a range of valid and invalid data. Under white-box (structure) testing, the challenge for the implementer is to ensure that the design has been correctly implemented and that the unit can be successfully traversed through each of its decision paths.

Implementation: Test Unit within a Subsystem

Unit test focuses on verifying the smallest testable components of the software. Unit tests are designed, implemented, and executed by the implementer of the unit. The emphasis of unit test is to ensure that white-box testing produces the expected results and that the unit conforms to the project's adopted quality and development standards.

Implementation and Test: Integrate a Subsystem

The purpose of subsystem integration is to combine units that may come from many different developers within the subsystem (part of the implementation model) into an executable build set. In accordance with the plan, the implementer integrates the subsystem by bringing together completed and stubbed classes that constitute a build. The implementer integrates the subsystem incrementally from the bottom up based on the compilation dependency hierarchy.

Implementation: Test a Subsystem

Testers execute test procedures developed earlier. If there are any unexpected test results, integration testers log the defects for arbitration about when they are to be fixed.

Implementation: Release a Subsystem

Once the subsystem has been sufficiently tested and it is ready for integration at the system level, the implementer *releases* the tested version of the subsystem from the team integration area into an area where it becomes visible, and usable, for system-level integration.

Implementation: Integrate the System

The purpose of system integration is to combine the currently available implementation model functionality into a build. The system integrator incrementally adds subsystems and creates a build that is handed over to the integration testers for overall integration testing.

Test: Test Integration

Integration testers execute test procedures developed earlier. The integration testers execute integration tests and review the results. If there are any unexpected results, the integration testers log the defects.

Test: Test the System

Once the whole system (as defined by the goal of this iteration) has been integrated, the system tester tests the system. The test designer then analyzes the results of the test to make sure that the testing goals have been reached.

Project Management: Assess the Iteration Itself

The project manager compares the iteration's actual cost, schedule, and content with the iteration plan; determines whether rework needs to be done and, if so, assigns it to future iterations; updates the risk list (Artifact: Risk List); updates the project plan (Artifact: Project Plan); and prepares an outline of an iteration plan for the next iteration (Artifact: Iteration Plan).

Result

The main result of a late iteration in the construction phase is that more functionality is added, and that yields an increasingly more complete system. The results of the current iteration are made visible to developers to form the basis of development for the subsequent iterations.

SUMMARY

- Each iteration runs through each of the core workflows.
- Whether an activity is visited or revisited depends in large part on which phase in the life cycle the project currently is in and on the artifacts that the development case prescribes for this project.

Chapter 17

Configuring and Implementing the Rational Unified Process

with Håkan Dyrhage

THIS CHAPTER DESCRIBES the strategies and tactics for configuring and implementing the Rational Unified Process in an adopting software development organization.

INTRODUCTION

In many circumstances, the Rational Unified Process can be used in whole or in part "out of the box." Often, however, to do a thorough job and adapt it closely to the needs of your organization, you will have to configure and implement it.

To *configure* the Rational Unified Process means to adapt the process product to the needs and constraints of the adopting organization by modifying the process framework delivered by Rational Software.

To *implement* the Rational Unified Process in a software development organization means to change the organization's practice so that it routinely and successfully uses the Rational Unified Process in whole or in part.

The result of configuring the Rational Unified Process is captured in a *development case*. A development case could constitute an enumeration of the different changes to be made to the complete Rational Unified Process. Alternatively, the development case could be a Web site with online hyperlinks to the relevant sections of the Rational Unified Process.

THE EFFECT OF IMPLEMENTING A PROCESS

Process changes are difficult, and it may be a long time before you see their true effects. This is in contrast to the adoption of a new tool, something that is relatively easy and fast: you install it, read the user manual, go through an example, and maybe attend a training course. The transition to a new tool might last from a few hours to a couple of weeks. But changing the software development process often means affecting the fundamental beliefs and values of the individuals involved and changing the way they perceive their work and its value. It is a cultural change and often political or philosophical in nature.

A process change affects the individuals and the organization more deeply than a change of technology or tools. It must be carefully planned and managed. The adopting organization must identify the opportunity and the benefits, convey them clearly to the interested parties, raise their level of awareness, and then gradually change from the current practice to a new practice. Ivar Jacobson describes this as "reengineering your software engineering process."[1]

When implementing a process, you must address the following areas.

- *The people and their competence, skills, motivation, and attitude:* You must make sure that everyone is adequately trained and motivated.
- *The supporting tools:* You must buy new tools, replace old ones, and customize and integrate others.
- *The software development process:* This includes the lifecycle model, the organizational structure, the activities to be performed, the practices to be followed, the artifacts to

1. Ivar Jacobson and Sten Jacobson. "Reengineering Your Software Engineering Process," *Object Magazine*, March-April, 1995.

be produced, and the scope of the software development process.

- The *description* of the software development process.

Other areas also affect the way people work—for example, the physical working environment, the organization's culture and politics, and the reward structure. Later in this chapter we'll touch on some of these issues.

In addition to the people whose work is most directly affected by the process change, you must take into consideration the people who otherwise might feel excluded from the change process.

- Managers are responsible for the performance of the software development organization. They must understand why you are changing the process and why you are procuring new tools. It is important that they understand how (and whether) progress is being made. Any process improvement project must have executive support. You must make sure that management understands that there is a return on the investment being made in changing the process, and you must also manage that expectation carefully.
- Customers must be informed that your process effectively takes their needs into consideration to evolve the product as their needs change.
- Other parts of the software development organization are also affected. Sometimes you change only one part of the organization, and that may lead to resistance and skepticism from other parts of the organization. Often, the people in other departments do not understand what you are doing and why you are doing it. Even if they do not have a direct influence, their exclusion may cause political problems.

INVOLVE PEOPLE

To succeed with the process-implementation project, it is important to involve people in the effort as early as possible. They are an important source of information when you are assessing the software development organization's current state. Second, you should ensure that everyone understands the current state of the organization and perceives the problems they're experiencing as well as how and where they can improve. Building up this understanding is one of the keys to success for any change project.

Start early. Let the entire organization know what is going on and where the process change is heading. There is always a potential risk that people outside the affected part of the organization will not support your effort. One way to reduce this risk is to keep them informed. Communicating this information is important, and you should allow enough time for it.

Make sure that you have chosen the right people to participate in the first part of the change project because they will be the ones who will pass on the message to the rest of the organization. Some of them will play important roles as mentors when the process is implemented in the rest of the organization.

IMPLEMENTING THE RATIONAL UNIFIED PROCESS STEP-BY-STEP

Implementing a new process in a software development organization can be described in six steps (see Figure 17-1).

Step 1: Assess the Current State

You need to understand the current state of the software development organization in terms of its people, process, and supporting tools. You should also identify problems and potential areas for improvement as well as collect information about outside issues such

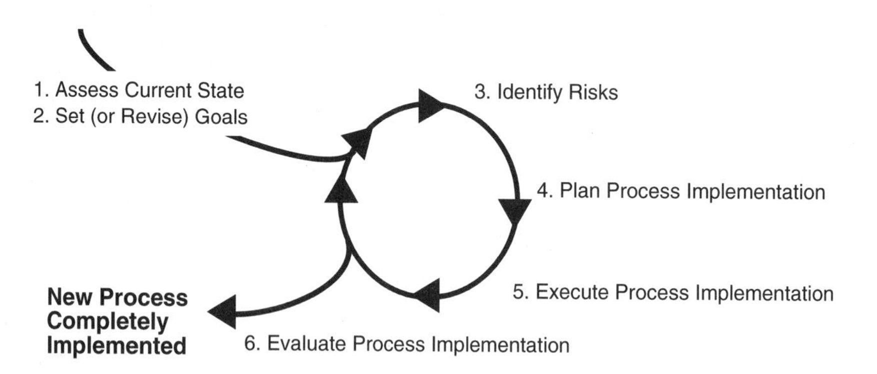

FIGURE 17-1 *The steps to implement a new software development process*

as competitors and market trends. When this step is complete, you should know the following:

- The current state of the software development organization
- The kind of people who work here and their level of competence, skills, and motivation
- The tools currently used in the organization
- The current software engineering process and how it is described

Why should you assess the current state? Consider the following reasons.

- You want to use it to create a plan for getting from the current state of the organization to your goal.
- You want to identify the areas that need to be improved first. You may not want to introduce the entire process and all the tools at once. Instead, you may prefer to do it in increments, starting with the areas that have the greatest need and the best potential for improvement.
- You want to explain to the sponsors why you need to change the process, tools, and people.
- You want to create motivation and a common understanding among the people who are directly or indirectly affected.

It is important to have an understanding of the project's level of management complexity and level of technical complexity. The more stakeholders the project has and the bigger the project, the higher the level of ceremony that is needed. More artifacts must be produced, communicated, explained, reviewed, and approved. The higher the level of technical complexity, the more effort that must be put into maintaining the artifacts and tracking the status of the project (see Figure 17-2).

Step 2: Set (or Revise) Goals

The second step is to set up goals for the process, people, and tools, noting where you want to be when the process implementation project is complete. You set up goals for the following reasons.

- Goals serve as an important input to planning the implementation of the process.

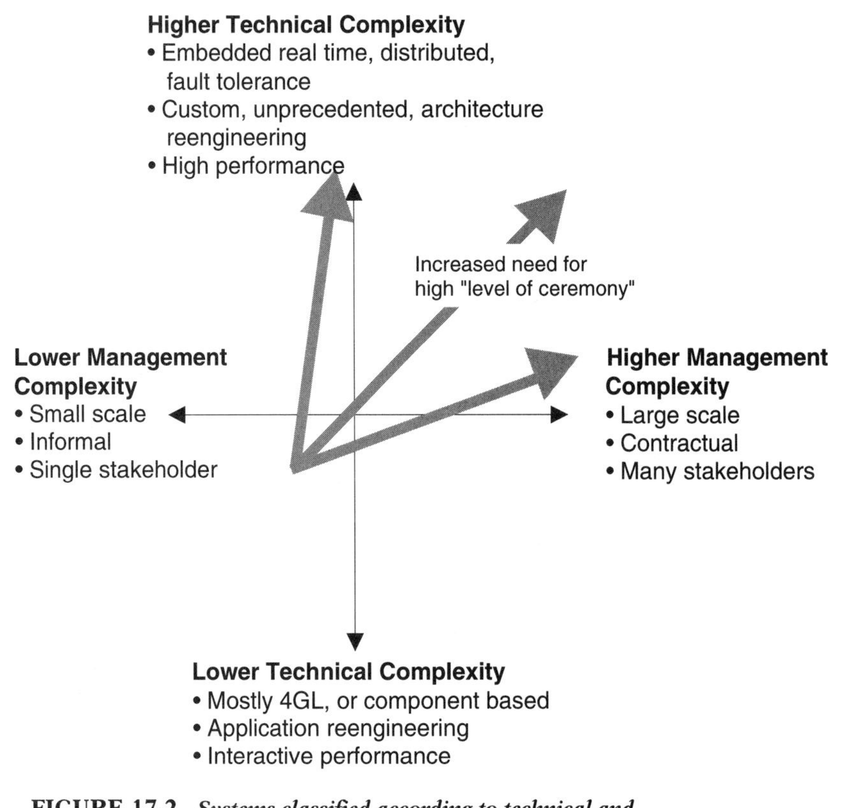

FIGURE 17-2 *Systems classified according to technical and managerial complexity*

- Goals, combined with the result of step 1 (a description of the current state), are used to motivate the sponsors and to create an understanding and motivation among the people in the organization.

The result should be a list of measurable goals that are expressed so that they can be comprehended and internalized by project members. The goals can serve as a vision of the future state of the organization.

Step 3: Identify Risks

To succeed in implementing a new process, you must control the many risks involved. We recommend that you perform a risk analysis in which you identify potential risks, try to understand their impact, and then rank them. You should also plan how you will mitigate the risks and in what order (as described in Chapter 7).

Switching from a linear process to an iterative process is not a risk-free undertaking. Software development organizations using an iterative approach for the first time may fall into several traps, including the following.

- The first iteration is too ambitious, and people cannot focus on the most important risks.
- You fail to complete the iterations, thinking that you have mitigated a risk without having tested what you built.
- Not all stakeholders buy into or understand the iterative approach, and they expect the project to progress as if it were using the waterfall method.
- You plan feature-by-feature instead of planning the complete project from the start (at a high level) and being prepared to change the plan along the way.
- There is too much rework between the iterations, meaning that you do things in an earlier iteration that you must redo later.
- You fail to control requirements changes, thinking that changes do not matter very much because problems can be fixed in later iterations.

Other process-related risks include the following.

- You undertake too much training too early. As a result, too much time elapses before people start to produce real work and they tend to forget what they learned earlier.
- You staff up the project too quickly, and too many people are involved in the inception phase. This makes it difficult for workers to understand their responsibilities and runs the risk of leaving some people without tasks.
- Tool support is insufficient, or people lack the skills to properly use them.
- You fail to assess the true value of artifacts and thus produce unnecessary artifacts.

Step 4: Plan the Process Implementation

You should develop a plan for implementing the process and the tools in the organization. This plan should describe how to efficiently move from the current state of the organization to the goal state.

Do not try to do everything at once. Instead, divide the implementation into a number of increments, and for each one, implement a portion of the new process together with the supporting tools. Typically, you should focus on one of the areas where you believe the change will have the most impact. If the software development organization is weak in testing, you might start by introducing the test workflow of the Rational Unified Process together with tools to automate testing. If, on another hand, the organization is weak in capturing or managing requirements, you might start by introducing the requirements workflow together with the supporting tools.

There are different approaches for implementing a process. The one you choose depends on two things.

- *The need for change in the current organization*
  If there are a lot of problems in the organization—problems with the tools or with the way people work—the frustration level in the organization will be high. In this case, you can be more aggressive and employ the new process and tools, or parts of them, in real projects.
- *The risks involved*
  If the risks are small, you can be more aggressive and can more quickly start using the process and tools in new projects. If the risks are great you must be more careful, using pilot projects to verify the process and the tools.

We give examples of two approaches:

- The *"typical"* approach
- The *"fast"* approach

In the typical approach (illustrated in Figure 17-3), you first implement the process in a pilot project. In an initial step you configure the process and describe it in a development case. You use the process on the pilot project. Experience from the pilot project is fed back into the development case. The process is then considered

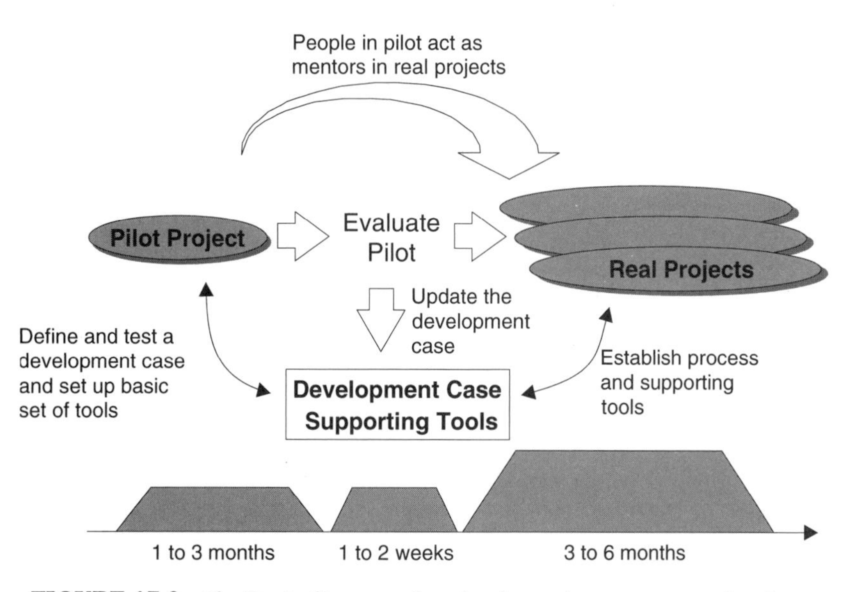

FIGURE 17-3 *The "typical" approach to implementing a process and tools*

tested, or verified, and can be rolled out to a broader audience. As a pilot project you might use one of the following:

- A complete software development project that is considered to have low risk from a technical and financial perspective
- The first complete iteration of an actual software development project, with the caveat that the main focus is on learning and improving the process and not on developing software

This approach is the most effective way to introduce the process and tools.

The fast approach (illustrated in Figure 17-4) is to use the process and tools directly in actual projects without first verifying that they work in a pilot project. This approach introduces a greater risk of failure, but there can be good reasons for taking these risks. For example, if the current process is very similar to the Rational Unified Process and if the tools are already used in the organization, it may be relatively easy and low-risk to implement the new process and tools.

Another case is an organization that is suffering from such severe problems that any change is perceived as an improvement.

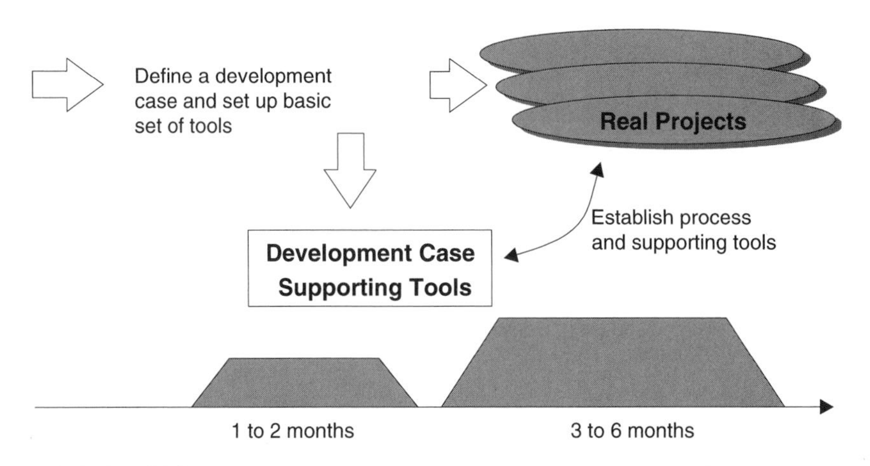

FIGURE 17-4 *The "fast" approach to implementing a process and tools*

This assumes that the potential for improvement is greater than the problems the organization inevitably will encounter.

Step 5: Execute the Process Implementation

The most time-consuming step in implementing a process is to execute it according to the plan defined in step 4. This step includes the following tasks.

- Develop a new development case or update an existing one.
- Acquire and adapt tools to support and to automate the process.
- Train members of the development team to use the new process and tools.
- Apply the process and tools in a software development project.

Step 6: Evaluate the Process Implementation

When you have implemented the process and tools in a software development project, actual or pilot, you must evaluate the effort. Did you achieve the goals you established? Evaluate the people, the process, and the tools to understand which areas you should focus on when you start again from step 1.

CONFIGURING THE PROCESS

In general, the software engineering process can be adapted or modified on two levels.

- *The company-process level*
  Process engineers modify, improve, or tailor a common process to be used companywide. This approach takes into consideration issues such as the domain of the application, reuse practices, and key technologies mastered by the company. One company can have more than one company process, each one adapted for a different type of development.
- *The project-process level*
  Process engineers analyze the company-level process to refine it for a given project. This level takes into consideration the size of the project, the reuse of company assets, initial cycle ("greenfield development") versus evolution cycle, and so on. The project-process level is the level you normally describe in a development case.

The project manager operates in the framework defined by the project-level process. The project managers make decisions about the pragmatic tasks: the milestones, the number and duration of iterations, the artifacts to be produced, and staffing (see Chapter 7).

Modification of the Rational Unified Process Product

In some cases, there is a need to modify the online version of the Rational Unified Process and thereby configure the process. After putting a baseline copy of the Rational Unified Process online under configuration management, process engineers modify it to incorporate changes such as the following:

- Add, expand, modify, or remove steps in activities
- Add checkpoints to the review activities based on experience, especially for problems discovered late in the development cycle
- Add guidelines, also based on discoveries made in past projects
- Tailor the templates: add company logo, header and footer, identification, and cover page
- Add tool mentors as needed

Some changes, however, are harder than others:

- Changes in process terminology that have a sweeping effect
- Using a process model different from the one presented in Chapter 3
- Changing the core workflow structure

The amount of work to create the corresponding development case may be considerable, and the reconciliation with Rational's future releases of the Rational Unified Process may be more difficult.

IMPLEMENTING A PROCESS IS A PROJECT

Implementing a software development process is a complex task that should be carefully controlled. We recommend treating it as a project (external to or a subproject of your software development project) and setting up milestones, allocating resources, and managing it as you would for any project.

The process-implementation project is divided into a number of phases. All six steps are performed in each phase until the project is ready and the process and tools are deployed and successfully used by the entire organization (see Figure 17-5). Table 17-1 summarizes how a project can be planned with four phases.

The group of people working on implementing the process should be dedicated to this task. They should function as mentors in the software development project, applying the process and tools. It is also their responsibility to maintain the new process, and that includes incorporating improvements suggested by the users. Also, to make sure that the process gains credibility within the organization, this group must make clear to the rest of the organization the impact of the new process on productivity and product quality. In large organizations, you may have a process engineering

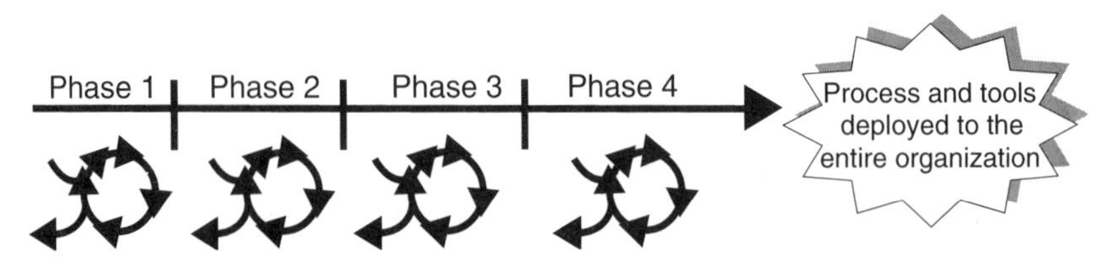

FIGURE 17-5 *A process-implementation project divided into phases*

team (such as the Software Engineering Process Group, SEPG) that configures and maintains the process as well as toolsmiths who adapt and maintain the supporting tools (see Figure 17-6).

TABLE 17-1 *The Four Phases of a Process-Implementation Project*

Phase	Purpose	Important Results after the Phase
1	To "sell" the process-implementation project to the sponsors	Go/no-go decision from the sponsors. To support the decision, the tools may be demonstrated and a development case may be exemplified.
2	To handle the major risks	Some tools ready to use; critical parts of development case ready.
3	To complete everything	All tools are ready, the development case is complete, a training curriculum is ready; mentors are ready to support real projects in next phase.
4	To deploy it to the entire organization	Process and tools are deployed to the entire organization.

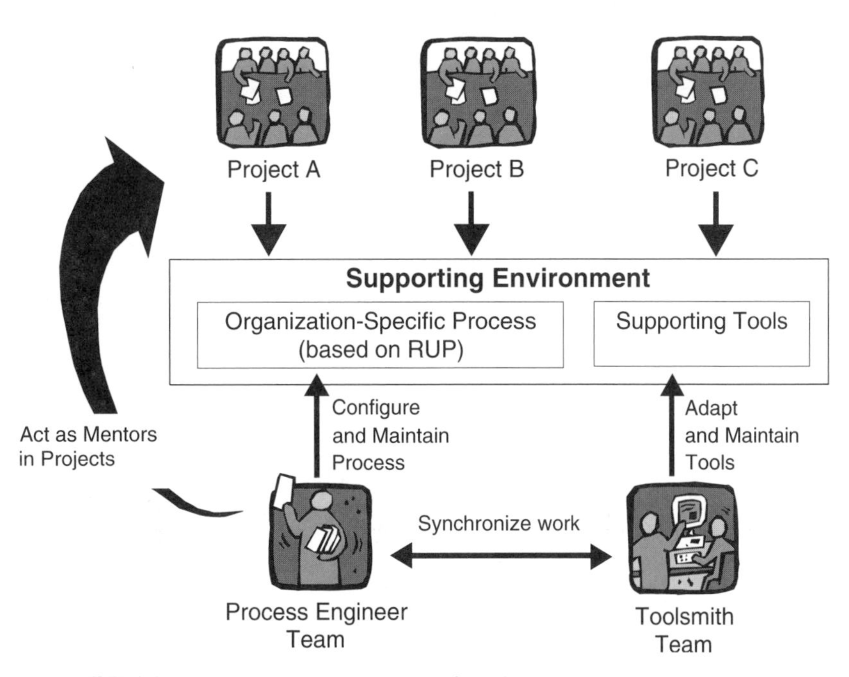

FIGURE 17-6 *Process engineers and toolsmiths*

SUMMARY

- An organization can take various approaches to adopt the Rational Unified Process.
- The typical approach involves trying part of the Rational Unified Process in a pilot project before extending it to the entire organization.
- Adopting the Rational Unified Process usually involves developing a development case, which is a project-specific version of the process.
- The development case is likely to be a Web site. It can be a modification of the Rational Unified Process online or a Web site referring to the Rational Unified Process online by way of hyperlinks.
- Achieving a process change is difficult, because it is a cultural change and is not without inherent risks.
- Tools to support the process must be put in place simultaneously with the process.
- Professional education and tool training are an integral part of a carefully planned process change.

Appendix A
Summary of Workers

This appendix summarizes the workers involved in the Rational Unified Process. Remember that a worker is not an individual but rather represents the roles that individuals play and their responsibilities for artifacts (see Chapter 3). They are listed in alphabetical order.

Architect The architect leads and coordinates technical activities and artifacts throughout the project. The architect establishes the overall structure for each architectural view: the decomposition of the view, the grouping of elements, and the interfaces between these major groupings.

Architecture Reviewer The architecture reviewer plans and conducts the formal reviews of the software architecture in general.

Business Designer The business designer details the specification of a part of the organization by describing the workflow of one or several business use cases. This worker details the specification of a part of the organization by describing the workflow of one or several business use cases. He or she specifies the business workers and business entities needed to realize a business use case and distributes the behavior of the business use case to them. The business designer defines the responsibilities, operations, attributes, and relationships of one or several business workers and business entities.

Business-Model Reviewer The business-model reviewer participates in formal reviews of the business use-case model and business object model.

Business-Process Analyst The business-process analyst leads and coordinates business use-case modeling by outlining and delimiting the organization that is being modeled.

Code Reviewer A code reviewer is responsible for ensuring the quality of the source code and for planning and conducting source code reviews.

Configuration Manager A configuration manager is responsible for providing the overall CM infrastructure and environment for the product development team. The CM function supports the product development activity so that developers and integrators have appropriate workspaces to build and test their work and that all artifacts are available as required. The configuration manager ensures that the CM environment facilitates product review, change, and defect tracking activities. The configuration manager is also responsible for writing the CM plan and reporting change-request-based progress statistics.

Course Developer The course developer develops training material to teach users how to use the product. He or she creates slides, student notes, examples, tutorials, and so on to enhance the understanding of the product.

Database Designer The database designer defines the tables, indexes, views, constraints, triggers, stored procedures, tablespaces or storage parameters, and other database-specific constructs needed to store, retrieve, and delete persistent objects.

Deployment Manager The deployment manager is responsible for plans to transition the product to the user community. These tasks are documented in deployment plans.

Design Reviewer The design reviewer plans and conducts the formal reviews of the design model.

Designer The designer defines the responsibilities, operations, attributes, and relationships of one or several classes and determines how they should be adjusted to the implementation environment. In addition, the designer may have responsibility for one or more design packages or design subsystems, including any classes owned by the packages or subsystems.

Implementer An implementer is responsible for developing and testing components in accordance with the project's adopted standards so that they can be integrated into larger subsystems. When test components, such as drivers or stubs, must be created to support testing, the implementer is also responsible for developing and testing the test components and corresponding subsystems.

Integration Tester The integration tester is responsible for executing the integration tests.

Performance Tester The performance tester is responsible for executing the performance tests.

Process Engineer The process engineer is responsible for the software development process itself. This includes configuring the process before project start-up and continuously improving the process during the development effort.

Project Manager The project manager allocates resources, shapes priorities, coordinates interactions with the customers and users, and generally tries to keep the project team focused on the right goal. The project manager establishes a set of practices that ensure the integrity and quality of project artifacts. The project manager is also responsible for ensuring that there is an effective product change review process.

Requirements Reviewer The requirements reviewer plans and conducts the formal review of the use-case model.

System Administrator The system administrator maintains the development environment—both hardware and software—and is responsible for system administration, backup, and so on.

System Analyst The system analyst leads and coordinates requirements elicitation and use-case modeling by outlining the system's functionality and delimiting the system.

System Integrator Implementers promote their tested components into an integration area, where system integrators combine them to produce an internally released build. A system integrator is also responsible for planning the integration of the system.

System Tester The system tester is responsible for executing the system tests.

Technical Writer The technical writer produces end-user support material, such as user guides, help texts, release notes, and so on.

Test Designer The test designer is responsible for the planning, design, implementation, and evaluation of testing, including generation of the test plan and test model, implementation of the test procedures, and evaluation of test coverage, test results, and effectiveness.

Toolsmith The toolsmith develops tools to support special needs, to provide additional automation of tedious or error-prone tasks, and to provide better integration between tools.

Use-Case Specifier The use-case specifier details the specification of a part of the system's functionality by describing the requirements aspect of one or several use cases. The use-case specifier can also be responsible for a use-case package and for maintaining the integrity of that package. The use-case specifier responsible for a use-case package is also responsible for its contained use cases and actors.

User-Interface Designer The user-interface designer leads and coordinates the prototyping and design of the user interface by capturing requirements on the user interface, including usability requirements; building user-interface prototypes; involving other stakeholders of the user interface, such as end users, in usability reviews and use testing sessions; and reviewing and providing the appropriate feedback on the final implementation of the user interface.

Appendix B
Summary of Artifacts

This appendix summarizes the artifacts that are produced or used during the course of the process. They are organized according to the five major information sets (see Chapter 3) and are associated with the responsible worker.

The Requirements Set

Artifact	Worker
■ Vision	System analyst
■ Stakeholders' needs	System analyst
■ Use-case model	System analyst
– Use case	Use-case specifier
– Actor	System analyst
– Use-case package	Use-case specifier
■ Supplementary specification	System analyst
■ Glossary: System analyst	
■ User-interface prototype	User-interface designer
■ Use-case storyboard	User-interface designer

If business modeling is performed:

Artifact	Worker
■ Business use-case model	Business-process analyst
– Business use case	Business designer
– Business worker	Business designer
■ Business object model	Business-process analyst
– Business entity	Business designer
– Organization unit	Business designer
■ Supplementary business specification	Business-process analyst

The Design Set

Artifact	Worker
■ Software architecture description	Architect
■ Design model	Architect
– Class	Designer

Artifact	Worker
Design model *continued*	
– Package	Designer
– Subsystem	Designer
– Interface	Designer
■ Test model	Test designer
– Test case	Test designer
– Test procedure	Test designer
– Test script	Test designer
– Workload model	Test designer
■ Analysis model	Architect
■ Data model	Database designer
■ Deployment model	Architect
■ Process model	Architect

The Implementation Set

Artifact	Worker
■ Implementation model	Architect
– Source code	Implementer
– Associated files	Implementer
– Executable files	Implementer
– Test code	Implementer

The Deployment Set

Artifact	Worker
■ Executable baselines	Implementer
■ Release notes	Technical writer
■ User support material	Technical writer
– User manual	Technical writer
– Installation material	Technical writer
– Training material	Course developer

The Management Set

Artifact	Worker
■ Business case	Project manager
■ Software development plan	Project manager
– Development case	Process engineer
• Guidelines (design, test, etc.)	Many workers
– Project plan	Project manager
– Configuration management plan	Configuration manager
– Measurement plan	Project manager
– Risk list	Project manager
■ Iteration plans	Project manager
■ Iteration assessment	Project manager
■ Status assessment	Project manager
■ Change request	Project manager
■ Deployment plan	Deployment manager

Glossary

abstraction The essential characteristics of an entity that distinguish it from all other kinds of entities and thus provide crisply defined boundaries relative to the perspective of the viewer.

activity A unit of work that a worker may be asked to perform.

actor (instance) Someone or something outside the system or business that interacts with the system or business.

actor class A class that defines a set of actor instances in which each actor instance plays the same role in relation to the system or business.

architectural baseline The baseline at the end of the elaboration phase, at which time the foundation structure and behavior of the system are stabilized.

architectural pattern A description of an archetypal solution to a recurrent design problem that reflects well-proven design experience.

architectural view A view of the system architecture from a given perspective; focuses primarily on structure, modularity, essential components, and the main control flows. See also *view*.

architecture See *software architecture*.

artifact A piece of information that is produced, modified, or used by a process, defines an area of responsibility, and is subject to version control. An artifact can be a model, a model element, or a document.

baseline A reviewed and approved release of artifacts that constitutes an agreed-on basis for further evolution or development and that can be changed only through a formal procedure, such as change and configuration control.

build An operational version of a system or part of a system that demonstrates a subset of the capabilities to be provided in the final product.

change control board (CCB) The role of the CCB is to provide a central control mechanism to ensure that every change request is properly considered, authorized, and coordinated.

change management The activity of controlling and tracking changes to artifacts.

change request (CR) A request to change an artifact or process. Documented in the CR is information on the origin and impact of the current problem, the proposed solution, and its cost.

class A description of a set of objects that share the same responsibilities, relationships, operations, attributes, and semantics.

component A nontrivial, nearly independent, and replaceable part of a system that fulfills a clear function in the context of a well-defined architecture. A component conforms to and provides the physical realization of a set of interfaces.

component-based development (CBD) The creation and deployment of software-intensive systems assembled from components as well as the development and harvesting of such components.

configuration (1) General: The arrangement of a system or network as defined by the nature, number, and chief characteristics of its functional units; applies to both hardware and software configuration. (2) The requirements, design, and implementation that define a particular version of a system or system component.

configuration management (CM) A supporting process workflow whose purpose is to identify, define, and baseline items; control modifications and releases of these items; report and record status of the items and modification requests; ensure completeness, consistency, and correctness of the items; and control storage, handling, and delivery of the items. (ISO)

construction The third phase of the Rational Unified Process, in which the software is brought from an executable architectural baseline to the point at which it is ready to be transitioned to the user community.

cycle One complete pass through the four phases: inception, elaboration, construction, and transition. The span of time between the beginning of the inception phase and the end of the transition phase.

defect A product anomaly. Examples include omissions and imperfections found during early life-cycle phases and symptoms of faults contained in software sufficiently mature for test or operation. A defect can be any kind of issue you want tracked and resolved.

deliverable An output from a process that has a value, material or otherwise, to a customer.

deployment A core process workflow in the software engineering process whose purpose is to ensure a successful transition of the developed system to its users. Included are artifacts such as training materials and installation procedures.

deployment view An architectural view that describes one or several system configurations; the mapping of software components (tasks, modules) to the computing nodes in these configurations.

design The part of the software development process whose primary purpose is to decide how the system will be implemented. During design, strategic and tactical decisions are made to meet the required functional and quality requirements of a system.

design model An object model describing the realization of use cases; serves as an abstraction of the implementation model and its source code.

development case The software engineering process used by the performing organization. It is developed as a configuration or customization of the Rational Unified Process product and adapted to the project's needs.

domain An area of knowledge or activity characterized by a family of related systems.

elaboration The second phase of the process, in which the product vision and its architecture are defined.

environment A core supporting workflow in the software engineering process whose purpose is to define and manage the environment in which the system is being developed. Includes process descriptions, configuration management, and development tools.

evolution The life of the software after its initial development cycle; any subsequent cycle during which the product evolves.

framework A micro-architecture that provides an incomplete template for applications within a specific domain.

implementation A core process workflow in the software engineering process whose purpose is to implement and unit-test the classes.

implementation model A collection of components and the implementation subsystems that contain them.

implementation subsystem A collection of components and other implementation subsystems. Used to structure the implementation model by dividing it into smaller parts.

implementation view An architectural view that describes the organization of the static software elements (code, data, and other accompanying artifacts) on the development environment, in terms of packaging, layering, and configuration management (ownership, release strategy, and so on). In the Rational Unified Process, it is a view on the implementation model.

inception The first phase of the Rational Unified Process, in which the seed idea, or request for proposal, for the previous generation is brought to the point (at least internally) of being funded to enter the elaboration phase.

increment The difference (delta) between two releases at the end of subsequent iterations.

integration The software development activity in which separate software components are combined into an executable whole.

iteration A distinct sequence of activities with a baselined plan and valuation criteria resulting in a release (internal or external).

layer A specific way of grouping packages in a model at the same level of abstraction.

logical view An architectural view that describes the main classes in the design of the system: major business-related classes and the classes that define key behavioral and structural mechanisms (persistency, communications, fault tolerance, and user interface). In the Rational Unified Process, the logical view is a view of the design model.

management A supporting workflow in the software engineering process whose purpose is to plan and manage the development project.

method (1) A regular and systematic way of accomplishing something; the detailed, logically ordered plans or procedures followed to accomplish a task or attain a goal. (2) UML 1.2: The implementation of an operation; the algorithm or procedure that effects the results of an operation.

milestone The point at which an iteration formally ends; corresponds to a release point.

model A semantically closed abstraction of a system. In the Rational Unified Process, a complete description of a system from a perspective—"complete" meaning that you don't need any additional information to understand the system from that perspective; a set of model elements.

model element An element that is an abstraction drawn from the system being modeled.

node A runtime physical object that represents a computational resource, generally having at least a memory and often processing capability. Runtime objects and components may reside on nodes.

object An entity with a well-defined boundary and identity that encapsulates state and behavior. State is represented by attributes and relationships, and behavior is represented by operations and methods. An object is an instance of a class.

object model An abstraction of a system's implementation.

operation A service that can be requested from an object to effect behavior.

phase The time between two major project milestones during which a well-defined set of objectives is met, artifacts are completed, and decisions are made to move or not to move into the next phase.

prototype A release that is not necessarily subject to change management and configuration control.

quality The characteristic of having demonstrated the achievement of producing a product that meets or exceeds agreed-on requirements—as measured by agreed-on measures and criteria—and that is produced by an agreed-on process.

release A subset of the end product that is the object of evaluation at a major milestone.

report An automatically generated description that describes one or several artifacts. A report is not in itself an artifact. A report is in most cases a transitory product of the development process and a vehicle to communicate certain aspects of the evolving system; it is a snapshot description of artifacts that are not themselves documents.

requirement A description of a condition or capability of a system; either derived directly from user needs or stated in a contract, standard, specification, or other formally imposed document.

requirements workflow A core process workflow in the software engineering process whose purpose is to define what the system should do. The most significant activity is to develop a use-case model.

risk An ongoing or impending concern that has a significant probability of adversely affecting the success of major milestones.

scenario A described use-case instance; a subset of a use case.

sequence diagram A diagram that describes a pattern of interaction among objects arranged in a chronological order; it shows the objects participating in the interaction by their "lifelines" and the messages that they send to one another.

software architecture Software architecture encompasses

- The significant decisions about the organization of a software system
- The selection of the structural elements and their interfaces by which the system is composed together with their behavior as specified in the collaboration among those elements
- The composition of the structural and behavioral elements into progressively larger subsystems
- The architectural style that guides this organization, these elements, and their interfaces, their collaborations, and their composition

Software architecture is concerned not only with structure and behavior but also with usage, functionality, performance, resilience, reuse, comprehensibility, economic and technology constraints and trade-offs, and aesthetic issues.

stakeholder Any person or representative of an organization who has a stake—a vested interest—in the outcome of a project or whose opinion must be accommodated. A stakeholder can be an end user, a purchaser, a contractor, a developer, or a project manager.

stub A dummy or skeletal implementation of a piece of code used temporarily to develop or test another piece of code that depends on it.

test A core process workflow in the software engineering process whose purpose is to integrate and test the system.

transition The fourth phase of the process in which the software is turned over to the user community.

use-case (class) A sequence of actions a system performs that yields an observable result of value to a particular actor. A use-case class contains all main, alternative, and exception flows of events related to producing the observable result of value. Technically, a use case is a class whose instances are scenarios.

use-case model A model of what the system is supposed to do and the system environment.

use-case realization A description of the way a particular use case is realized within the design model, in terms of collaborating objects.

use-case view An architectural view that describes how critical use cases are performed in the system, focusing on architecturally significant components (objects, tasks, nodes). In the Rational Unified Process, it is a view of the use-case model.

version A variant of an artifact; later versions of an artifact typically expand on earlier versions.

view A simplified description (an abstraction) of a model that is seen from a given perspective or vantage point and omits entities that are not relevant to this perspective. See also *architectural view*.

vision The user's or customer's view of the product to be developed.

worker Role figure to be played by individual members in the business.

workflow The sequence of activities performed in a business that produces a result of observable value to an individual actor of the business.

Bibliography

This highly selective bibliography records the favorite books (and a few articles) of the Rational Process Development group—books that have had a major impact on the Rational Unified Process as it stands today along with books that are complementary to the process and books published recently by our colleagues from Rational Software. Note that some books that could have appeared in several categories are listed only once.

General

Brooks, Frederick P., Jr. 1995. *The Mythical Man-Month Anniversary Edition: Essays on Software Engineering*. Reading, MA: Addison-Wesley.

A classic that should be read and reread by everyone involved in software development. I recommend this edition rather than the original 1975 edition.

Davis, Alan. 1995. *201 Principles of Software Development*. New York: McGraw-Hill.

Full of good advice for every worker.

Katzenbach, Jon R., and Douglas K. Smith. 1993. *The Wisdom of Teams*. New York: HarperBusiness.

The secret of effective teams.

Yourdon, Edward. 1997. *Death March: Managing "Mission Impossible" Projects*. Upper Saddle River, NJ: Prentice Hall.

An interesting view of project troubles.

Software Development Process

Boehm, Barry W. 1996. "Anchoring the Software Process." *IEEE Software*, July, pp. 73–82.

This article defines the four phases and the corresponding milestones.

Boehm, Barry W. 1998. "A Spiral Model of Software Development and Enhancement." *IEEE Computer*, May, pp. 61–72.

This seminal article defines the principles and motivations of iterative development.

Humphrey, W.S. 1989. *Managing the Software Process*. Reading, MA: Addison-Wesley.

A classic book on the software process and the capability maturity model developed at the Software Engineering Institute.

ISO/IEC 12207. 1995. *Information technology—Software life-cycle processes;* and ISO 9000-3. 1991. *Guidelines for the Application of ISO 9001 to the Development, Supply, and Maintenance of Software.* Geneva: ISO.

Two key standards for software process definition and assessment.

Jacobson, Ivar, Grady Booch, and James Rumbaugh. 1999. *The Unified Software Development Process.* Reading, MA: Addison Wesley Longman.

This recent textbook is a more thorough description of the Unified Process and is a useful companion to the Rational Unified Process. Also provides examples of UML modeling.

Jacobson, Ivar, Martin Griss, and Patrik Jonsson. 1997. *Software Reuse: Architecture, Process, and Organization for Business Success.* Reading, MA: Addison Wesley Longman.

This textbook on software reuse is great complement to the Rational Unified Process. It also features some great chapters on architecture.

Kruchten, Philippe. 1991. "Un processus de dévelopment de logiciel itératif et centré sur l'architecture." *Proceedings of the 4th International Conference on Software Engineering.* Toulouse, France, EC2, December.

The Rational iterative process explained in French.

Kruchten, Philippe. 1996. "A Rational Development Process." *CrossTalk* 9(7), pp. 11–16.

Developed with Walker Royce, Sue Mickel, and a score of Rational consultants, this article describes the iterative life cycle of the Rational Process.

McFeeley, Robert. 1996. *IDEAL: A User's Guide for Software Process Improvement.* Pittsburgh, PA: Software Engineering Institute.

This book describes a software process improvement program model: IDEAL, a generic description of a sequence of recommended steps for initiating and managing a process implementation project.

Paulk, Mark, et al. 1993. *Capability Maturity Model for Software, Version 1.1.* Pittsburgh, PA: Software Engineering Institute.

The original reference for the capability maturity model.

Object-Oriented Technology

Booch, Grady. 1994. *Object-Oriented Analysis and Design with Applications, Second Edition.* Menlo Park, CA: Addison-Wesley.

Jacobson, Ivar, et al. 1992. *Object-Oriented Software Engineering: A Use Case-Driven Approach,* Wokingham, UK: Addison-Wesley.

Rumbaugh, James, et al. 1991. *Object-Oriented Modeling and Design.* Englewood Cliffs, NJ: Prentice-Hall.

These three books, the original roots of the object-oriented analysis and design workflow, were written by "the three amigos" just before the advent of the UML and the Rational Unified Process. Despite the use of their original notations, these books are still the key references for the OO designer.

Buhr, R.J.A., and R.S. Casselman. 1996. *Use Case Maps for Object-Oriented Systems.* Upper Saddle River, NJ: Prentice-Hall.

This book develops other views on use cases.

Gamma, Erich, et al. 1995. *Design Patterns: Elements of Reusable Object-Oriented Software.* Reading, MA: Addison Wesley Longman.

Rumbaugh, James. 1996. *OMT Insights*. New York: SIGS Books.

A complement to the original OMT book, this one dives into special topics: inheritance, use cases, and so on.

Selic, Bran, Garth Gullekson, and Paul Ward. 1994. *Real-time Object Oriented Modeling*. New York: John Wiley & Sons.

The reference work on using object technology for the design of reactive systems by the people who brought us ObjecTime Developer.

Modeling and the Unified Modeling Language

Booch, G., J. Rumbaugh, and I. Jacobson. 1998. *The Unified Modeling Language*. Documentation set, version 1.3. Cupertino, CA: Rational Software.

The latest OMG standard on the UML at the time of this writing. Available online at www.rational.com.

Booch, Grady, James Rumbaugh, and Ivar Jacobson. 1999. *Unified Modeling Language Users Guide*. Reading, MA: Addison Wesley Longman.

Published at the same time as the Rational Unified Process 5.0, this manual is an excellent user's guide on UML by its main authors.

Eriksson, Hans-Erik, and M. Penker. 1997. *UML Toolkit*. New York: John Wiley & Sons.

A more comprehensive book on UML as seen from Sweden by another pair of Rational friends.

Fowler, Martin. 1997. *UML Distilled: Applying the Standard Object Modeling Language*. Reading, MA: Addison Wesley Longman.

A very nice little introduction to UML if you are in a hurry.

Muller, Pierre-Alain. 1998. *Instant UML*. Chicago: Wrox Inc.

Another short introduction to UML by a former colleague of ours.

Quatrani, Terry. 1998. *Visual Modeling with Rational Rose and UML*. Reading, MA: Addison Wesley Longman.

Provides step-by-step guidance on how to build UML models. At the same time, it follows the Rational Unified Process for which this book provides in effect a small-scale example.

Rumbaugh, James, Ivar Jacobson, and Grady Booch. 1999. *Unified Modeling Language Reference Manual*. Reading, MA: Addison Wesley Longman.

Certainly more digestible than the OMG standard, this book is an in-depth examination of UML by its main authors.

Project Management

Boehm, Barry W. 1991. "Software Risk Management: Principles and Practices." *IEEE Software*, Jan., pp. 32–41.

Still the best little introduction to risk management.

Booch, Grady. 1996. *Object Solutions: Managing the Object-Oriented Project*. Reading, MA: Addison-Wesley.

A pragmatic book for managers of object-oriented projects; one of the sources of the underlying philosophy of the Rational Unified Process.

Carr Marvin J., et al. 1993. *Taxonomy-Based Risk Identification*. Technical report CMU/SEI-93-TR-6, SEI, June.

A source of inspiration to get started on your own list of risks.

Charette, Robert. 1989. *Software Engineering Risk Analysis and Management*. New York: McGraw-Hill.

Practical perspective on risk management.

Fairley, Richard. 1994. "Risk Management for Software Project." *IEEE Software* 11(3), May, pp. 57–67.

Straightforward strategy for risk management if you have never done it before.

Gilb, Tom. 1988. *Principles of Software Engineering Management*. Harlow, UK: Addison-Wesley.

A great book by a pioneer of iterative development, full of pragmatic advice for the project manager.

Jones, Capers. 1994. *Assessment and Control of Software Risks*. Englewood Cliffs, NJ: Yourdon Press.

An indispensable source of risks to check to make sure your list is complete.

Karolak, Dale. 1996. *Software Engineering Risk Management*. Los Alamitos, CA: IEEE Computer Society Press.

Offers more sophisticated advice and techniques for risk management.

O'Connell, Fergus. 1994. *How to Run Successful Projects*. New York: Prentice-Hall International.

A real gem. Everything you really need to know to manage your first project, in 170 pages.

Royce, Walker. 1998. *Software Project Management: A Unified Framework*. Reading, MA: Addison Wesley Longman.

An indispensable companion to the Rational Unified Process, this book describes the spirit and the underlying software economics of the Rational Unified Process. It is full of great advice for the project manager.

Requirements Management

Davis, Alan. 1993. *Software Requirements: Objects, Functions and States*. Englewood Cliffs, NJ: Prentice Hall.

Gause, Donald, and Gerald Weinberg. 1989. *Exploring Requirements: Quality Before Design*. New York: Dorset House.

IEEE Std 830-1993. 1993. *Recommended Practice for Software Requirements Specifications*. New York: Software Engineering Standards Committee of the IEEE Computer Society.

Weinberg, Gerald. 1995. "Just Say No! Improving the Requirements Process." *American Programmer*, October.

Configuration Management

Berlack, H. 1992. *Software Configuration Management*. New York: John Wiley & Sons.

Buckley, J. 1993. *Implementing Configuration Management, Hardware, Software and Firmware*. Los Alamitos, CA: IEEE Computer Science Press.

Whitgift, David. 1991. *Methods and Tools for Software Configuration Management*. New York: John Wiley & Sons.

Testing and Quality

Beizer, Boris. 1995. *Black Box Testing*. New York: John Wiley & Sons.

A treasure of strategies to develop test cases for the functional testing of software. Dr. Beizer's writing style (and wit) makes this book easy and fun to read, with excellent, understandable examples.

Goglia, Patricia. 1993. *Testing Client/Server Applications*. Wellesley, MA: QED Press.

This was the first book that focused on testing client/server applications.

Hetzel, Bill. 1988. *The Complete Guide to Software Testing, Second Edition*. New York: J. Wiley/QED Press.

IEEE 829-1983. 1983. *Standard for Software Test Documentation*. New York: Software Engineering Standards Committee of the IEEE Computer Society.

Perry, William E. 1995. *Effective Methods for Software Testing*. New York: J. Wiley/QED Press.

Schmauch, Charles H. 1994. *ISO 9000 for Software Developers*. Milwaukee, WI: ASQC Quality Press.

Software Architecture

Buschmann, Frank, Régine Meunier, Hans Rohnert, Peter Sommerlad, and Michael Stahl. 1996. *Pattern-Oriented Software Architecture: A System of Patterns*. New York: John Wiley and Sons.

This book makes an inventory of a wide range of design patterns at the level of the architecture.

Bass, Len, Paul Clements, and Rick Kazman. 1998. *Software Architecture in Practice*. Reading, MA: Addison Wesley Longman.

A handbook of software architecture, with numerous case studies.

Hofmeister, Christine, Robert Nord, and Dilip Soni. 1999. *Applied Software Architecture*. Reading, MA: Addison Wesley Longman.

The architectural design approach these authors recommend is very similar to that of the Rational Unified Process and is based on multiple coordinated views.

IEEE Recommended Practice for Architectural Description. 1998. Draft 3.0 of IEEE P1471, May.

This proposed standard recommends architectural description based on the concept of multiple views.

Jacobson, Ivar, Martin Griss, and Patrik Jonsson. 1997. *Software Reuse: Architecture, Process and Organization for Business Success*. Reading, MA: Addison Wesley Longman.

A great companion book to the Rational Unified Process, this book offers insights into the design of components and of systems of interconnected systems. It lays out a strategy for institutionalizing the practice of systematic reuse at the corporate level.

Kruchten, Philippe. 1995. "The 4+1 view model of architecture." *IEEE Software* 12(6).

The origin of the 4+1 views used for architectural description in the Rational Unified Process.

Rechtin, Eberhardt. 1991. *Systems Architecting: Creating and Building Complex Systems*. Englewood Cliffs, NJ: Prentice-Hall; and Rechtin, Eberhard, and Mark Maier. 1997. *The Art of System Architecting*. Boca Raton, FL: CRC Press.

Although not specifically directed to software engineers, these two books are extremely valuable for software architects: they introduce an invaluable set of heuristics and many examples of architecture.

Shaw, Mary, and David Garlan. 1996. *Software Architecture: Perspectives on an Emerging Discipline.* Upper Saddle River, NJ: Prentice Hall.

A good introduction to the concepts and problems of software architecture.

Witt, Bernard I., F. Terry Baker, and Everett W. Merritt. 1994. *Software Architecture and Design: Principles, Models, and Methods.* New York: Van Nostrand Reinhold.

One of the first comprehensive books written on software architecture.

Business Engineering

Hammer, Michael, and James Champy. 1993. *Reengineering the Corporation: A Manifesto for Business Revolution.* New York: HarperBusiness.

The book that popularized the movement of business (re-)engineering. An excellent complement to the Jacobson book.

Jacobson, Ivar, Maria Ericsson, and Agneta Jacobson. 1994. *The Object Advantage: Business Process Reengineering with Object Technology.* Reading, MA: Addison-Wesley.

The basis of the business modeling workflow, this is the first book that applied object technology to the field of business engineering.

Others

DeGrace, Peter, and Leslie Stahl. 1990. *Wicked Problems, Righteous Solutions: A Catalog of Modern Software Engineering Practices.* Englewood Cliffs, NJ: Yourdon Press.

An insightful book on various process life cycles, their origins, their flaws, and their strengths; useful for an understanding of the importance of process.

Graham, Ian, et al. 1997. *The OPEN Process Specification.* Harlow, UK: Addison-Wesley.

Another process model, this one coming from down under, that shares some principles with the Rational Unified Process.

IBM. 1997. *Developing Object-oriented Software: An Experience Based Approach.* Upper Saddle River, NJ: Prentice Hall.

As with the Rational Unified Process, this book describes an iterative, incremental, object-oriented, scenario-driven, risk-aware process developed by IBM's Object Technology Center.

Kettani, Nasser, et al. 1998. *De Merise à UML.* Paris: Editions Eyrolles.

Merise, a popular software development methodology in France, has been upgraded to use UML. It has some similarity to the Rational Unified Process.

McCarthy, J. 1995. *Dynamics of Software Development.* Redmond, WA: Microsoft Press.

Fifty-three rules of thumb by a Microsoft development manager.

McConnell, Steve. 1993. *Code Complete: A Practical Handbook of Software Construction.* Redmond, WA: Microsoft Press.

A great book for the implementer and test workers, this one looks at the implementation, integration, and test aspects of the development process.

Stapleton, Jennifer. 1997. *The Dynamic System Development Method.* Reading, MA: Addison Wesley Longman.

At 15,000 feet, the DSDM approach could be seen as an introduction to the Rational Unified Process. Although they use a different terminology, the two processes are very close to each other, and you can see the Rational Unified Process as an instance or an implementation of DSDM.

Index

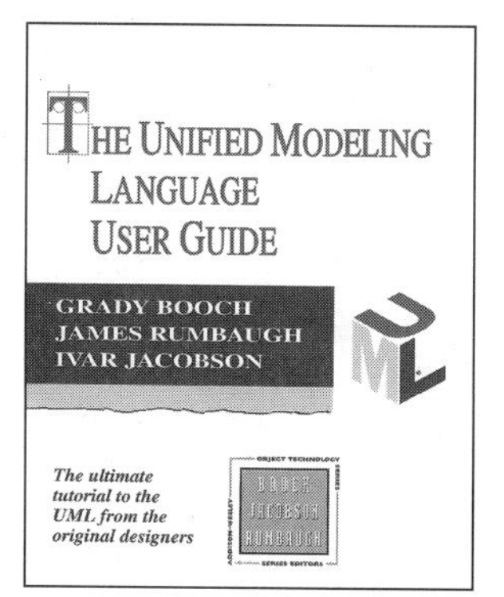

The Unified Modeling Language User Guide

Grady Booch, Ivar Jacobson, James Rumbaugh
Addison-Wesley Object Technology Series

The Unified Modeling Language User Guide is a two-color introduction to the core eighty percent of the Unified Modeling Language, approaching it in a layered fashion and showing the application of the UML to modeling problems across a wide variety of application domains. This landmark book is suitable for developers unfamiliar with the UML or modeling in general, but will also be useful to experienced developers who wish to learn how to apply the UML to advanced problems.

0-201-57168-4 • Hardcover • 512 Pages • ©1999

The Unified Software Development Process

Ivar Jacobson, Grady Booch, and James Rumbaugh
Addison-Wesley Object Technology Series

The Unified Software Development Process goes beyond other object-oriented analysis and design methods by detailing a family of processes that incorporate the complete life cycle of software development. This new book, representing the collaboration of Ivar Jacobson, Grady Booch, and James Rumbaugh, clearly describes the different higher-level constructs—notation as well as semantics—used in the models. Thus, stereotypes, such as use cases and actors, packages, classes, stereotypes, interfaces, active classes, processes and threads, nodes, and most relations, are described intuitively in the context of a model.

0-201-57169-2 • Hardcover • 512 pages • ©1999

The Unified Modeling Language Reference Manual

James Rumbaugh, Ivar Jacobson, and Grady Booch
Addison-Wesley Object Technology Series

James Rumbaugh, Ivar Jacobson, and Grady Booch have created the definitive reference to the Unified Modeling Language. This two-color book covers every aspect and detail of the UML and presents the modeling language in a useful reference format that serious software architects or programmers need to have on their bookshelves. The book is organized by topic and designed for quick access. The authors also provide the necessary information to enable existing OMT, Booch, and OOSE notation users to make the transition to the UML. This book provides an overview of the semantic foundation of the UML in a concise appendix.

0-201-30998-X • Hardcover with CD-ROM • 480 pages • ©1999